D1594698

HANDBOOK OF CONTROLS AND INSTRUMENTATION

John D. Lenk

Consulting Technical Writer

Prentice-Hall Inc., *Englewood Cliffs, New Jersey 07632*

Library of Congress Cataloging in Publication Data

Lenk, John D.
  Handbook of controls and instrumentation.
  Includes index.
  1. Process control. 2. Automatic control.
3. Engineering instruments. I. Title.
TS156.8.L45   629.8   79-25280

Editorial/production supervision and interior design: Nancy Moskowitz
Manufacturing buyer: Gordon Osbourne
Cover design: Maureen Olsen

Printed in the United States of America

10 9 8 7 6 5 4 3 2 1

PRENTICE-HALL INTERNATIONAL, INC., *London*
PRENTICE-HALL OF AUSTRALIA PTY. LIMITED, *Sydney*
PRENTICE-HALL OF CANADA, LTD., *Toronto*
PRENTICE-HALL OF INDIA PRIVATE LIMITED, *New Delhi*
PRENTICE-HALL OF JAPAN, INC., *Tokyo*
PRENTICE-HALL OF SOUTHEAST ASIA PTE. LTD., *Singapore*
WHITEHALL BOOKS LIMITED, WELLINGTON, *New Zealand*

To Irene, the "Sandpiper Lady,"
and Mr. Lamb, the "Magic Bunny"

CONTENTS

7 SUMMARY OF SENSING METHODS   *116*

8 MEASUREMENTS IN CONTROL AND INSTRUMENTATION SYSTEMS   *158*

9 BASIC CONTROL DEVICES   *212*

10 BASIC INSTRUMENTATION DEVICES   *280*

PREFACE

Our objective in this handbook is to describe basic operating principles of control and instrumentation systems. There are many kinds of control and instrumentation components, but the basic principles of the systems are similar for all types. For example, the fuel injection system of an automotive engine contains sensors to monitor engine and driving conditions such as compression, altitude, rotation, and speed. Information from those sensors is converted into control signals by a computer or computer-like device such as a microprocessor or controller. In turn, the signals are converted into commands that control the engine's performance; that is, commands regulate fuel air-gasoline mixture, amount of fuel, rate of fuel flow, etc. Thus, the engine is controlled or adjusted for optimum performance despite constantly changing operating conditions.

Similar components and techniques are applied to provide automatic control of industrial processes. Of course, the physical appearance of the sensors, controllers, and control devices is quite different for industrial and automotive applications, but the same basic principles of operation apply to all types of control and instrumentation devices and systems. For that reason, we concentrate on the basic principles of operation for controls and instrumentation in this book. Once you understand the basic, you can apply the principles to any system (industrial, automotive, aircraft, etc.).

Each application (industrial, automotive, missile control, etc.) has its own set of control and instrumentation systems suited to specific requirements. There are many variations of each system, and, obviously, it is impractical to discuss all systems in a single book. Instead, we consider a selected number of representative systems and devices. These are explained in simplified form to make clear their operating principles and to show that the basic principles apply to all systems and devices. An exhaustive investigation is not provided, since that would require technical background beyond that of the intended reader of this book. Thus, there are no complex mathematical formulas; the treatment is purely descriptive.

Since many control and instrumentation devices are electrical, electronic, mechanical, hydraulic, pneumatic, or a combination thereof, you should have some knowledge of electricity and electronics. Some background in mechanics, hydraulics, and pneumatics is also desirable but not essential.

Chapter 1 provides an introduction to control and instrumentation systems. Chapters 2 through 6 survey the sensors (or transducers) used in modern control and instrumentation systems, including sensors of motion, force, fluid, acceleration, attitude, displacement, torque, speed, velocity, strain, flow, pressure, liquid level, humidity, moisture, light, radioactivity, temperature, and sound. Chapter 7 summarizes sensing methods, and describes signal conditioning circuits used to convert signals from sensors into usable information. Chapter 7 also includes some specialized sensor systems used for measurement of thickness, proximity, density of liquids, specific gravity, and chemical content (acidity/alkalinity and gas analysis). Chapter 8 covers the role that measurements play in control and instrumentation systems. Chapters 9 and 10 concern basic control and instrumentation devices, respectively.

Many professionals have contributed their talent and knowledge to the preparation of this handbook. The author acknowledges that the effort to make this book a comprehensive work would be virtually impossible for one person, and he offers his appreciation to all who contributed to it, directly and indirectly.

The author also extends his gratitude to Jerry Slawney of Prentice-Hall, whose faith in this book has given the author encouragement, and also to take this opportunity to thank him for his expertise in making many of the author's books best sellers. The author also wishes to thank Mr. Joseph A. Labok of Los Angeles Valley College for his very special help with this and other books.

John D. Lenk

1

INTRODUCTION TO CONTROL AND INSTRUMENTATION SYSTEMS

Control and instrumentation systems can be very complex, or very simple, depending on the application. An example of a basic control system is a *thermostat,* which is used with many home heating units. The desired temperature is set into the system by a manual control. If the room temperature drops below the desired point, movement of a thermometer closes electrical contacts which, in turn, apply electrical power to a fuel valve. That power opens the valve, turns on a heater, and the room is heated. When the desired room temperature is reached, the thermometer opens the electrical contacts, removes power to the valve, and shuts down the heater.

1-1 TYPICAL INDUSTRIAL CONTROL AND INSTRUMENTATION SYSTEM

For an example of a more complex control and instrumentation system, consider an automated plant such as a petroleum refinery. The basic elements of such a system are shown in Fig. 1-1. Special devices called *transducers* or *sensors* (or possibly "pickups," although that is a slang term) are installed at many points along the refining process. The transducers measure the conditions or state of *process variables* at those

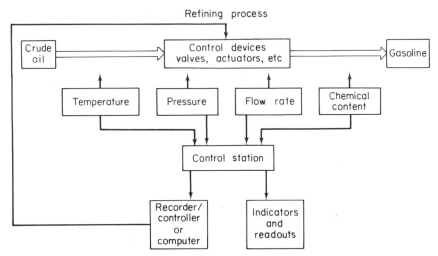

Refining process

Figure 1-1 Basic elements of a control and instrumentation system.

points. Typical process variables could include temperature, pressure, rate of flow, chemical content, density, etc., of the petroleum being refined. Any deviation from normal measurement (abnormal temperature, incorrect pressures, etc.) must be corrected instantly in order to keep the petroleum flowing without interruption and to ensure that the refinery output is of proper quality.

The output from each transducer or sensor is transduced, or converted, to a corresponding signal (usually electric but possibly pneumatic) and sent to a control center. That is, a process variable (temperature, pressure, etc.) is converted to a corresponding electrical signal in a form that can be accepted by a control station or device.

There are two basic types of control stations. A *recorder-controller* is used for simple systems. A recorder-controller prints a continuous record of the condition or state of a process variable and compares it with a normal or desired value (often called a *set point*) expected at that stage in a process. Any deviation from normal produces an output signal from the recorder-controller that corresponds to the *amount of deviation* as well as the *phase of deviation*. That is, the signal indicates whether the condition or state of the process variable is above or below the value indicated by the set point, and by how much. The recorder-controller output signal is sent to an *actuator* such as an electromagnetic valve or possibly a motor-driven device, which restores the process variable to its normal value.

In a large control system such as a petroleum refinery, there usually is a complex interrelationship among the process variables. This means that if one variable changes, that change usually affects all other variables. For example, an increase in the temperature of a liquid in a refining tank could

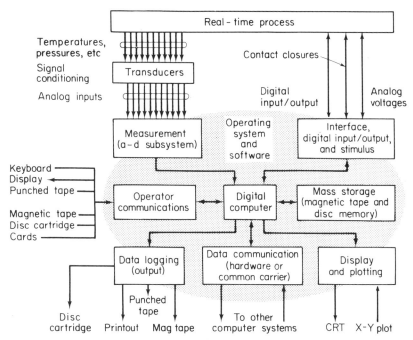

Figure 1-2 Block diagram of Hewlett-Packard 9600 automatic measurement and control system.

also raise the pressure on that liquid in the tank. Thus, if one variable changes, the set point of all the controllers should be readjusted to compensate for those changes. So complex is the interrelationship, however, that even an experienced operator cannot make all the readjustments required for the most efficient and profitable operation.

Digital computers, such as shown in Fig. 1-2, are used in the more complex control systems. These can be conventional larger computers or newer and smaller microcomputers or minicomputers. Whichever type of computer is used, signals from the transducers are converted to corresponding numerical quantities that are fed to the computer. That is, a physical quantity (pressure, temperature, etc.) is converted to an electrical signal by the transducer. In turn, an electrical signal is converted to a numerical quantity in a form suitable for the computer.

Because of its high speed, the computer is able to calculate the effect of every change in the interrelationships among the variables and quickly produce output signals that adjust the set point of each controller to produce the most efficient operation. In the more sophisticated systems, the computer usually replaces the older recorder-controller completely. In such advanced systems, the computer continuously prints out the condition or state of each process variable, stores the various set points in its memory, and ad-

3

justs the set points as required. The computer then compares the input signals from the transducers with the readjusted set points and sends appropriate signals to the various actuators.

1-2 TYPES OF CONTROL AND INSTRUMENTATION SYSTEMS

Each industry or specialized field has its own set of automatic control systems suited to its particular requirements, and there are many variations of each system. However, all control systems fall into either of two general groups: *open-loop* or *closed-loop* systems. Furthermore, these two basic groups can be classified as *continuous* or *discontinuous* systems. In the following sections we describe each of these four basic systems, followed by an explanation of a simple yet complete system that incorporates all the basic elements of a typical control system (including sensor, indicator, recorder, controller, actuator, etc.).

1-3 BASIC DISCONTINUOUS OPEN-LOOP CONTROL SYSTEM

In any open-loop system, a controlling device operates independently of a process variable that it controls. An example of a discontinuous open-loop system is shown in Fig. 1-3. The switch is the controlling device. The room temperature is the "process variable." When the switch is closed, current flowing through the heater raises the room temperature. When the switch is opened, the room temperature starts to fall. Because operation of the switch is not affected by room temperature (or the condition of the process variable) the system is open-loop. Likewise, because the control circuit can be opened and closed, the system is discontinuous.

1-4 BASIC CONTINUOUS OPEN-LOOP CONTROL SYSTEM

A basic continuous open-loop system is shown in Fig. 1-4. Here the rheostat (or variable resistor) is the controlling device. The resistance of the rheostat controls the amount of current flowing through the heater. In turn, the amount of current determines the room temperature. Again, the condition of the process variable (or room temperature) does not affect operation of the controlling device (rheostat), so the system is open-loop. However, because the control circuit is always closed and in operation, the system is continuous.

In most continuous open-loop systems, a *calibrated controlling device* must be used to place the process variable at the desired value. In the example shown in Fig. 1-4, the dial which operates the rheostat could be

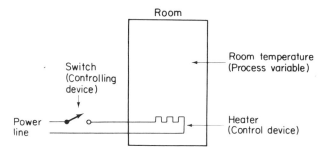

Figure 1-3 Basic discontinuous open-loop control system.

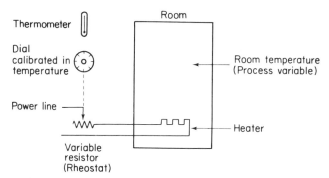

Figure 1-4 Basic continuous open-loop control system.

calibrated in terms of desired room temperature. However, variations such as opening and closing windows and doors to the room affect such calibrations. From a practical standpoint, the rheostat has no means for sensing such changes and making adjustments to meet them.

In most control systems, it is usually desired that the process variable be kept at a constant value. If an open-loop system such as shown in Fig. 1-4 is used, an *indicating device* such as a thermometer must be used to show the condition of the process variable (room temperature) at all times. Then an operator must be posted at the controlling device to decrease the rheostat resistance (thus increasing current flow and heat) should the room temperature fall below the desired value (set point) or to increase the rheostat resistance should the temperature rise above that value.

1-5 BASIC DISCONTINUOUS CLOSED-LOOP CONTROL SYSTEM

An obvious disadvantage of an open-loop system is the lack of *automatic control*. That is, the process variable (or room temperature in our case) is not monitored constantly and is not maintained at a desired value, except by

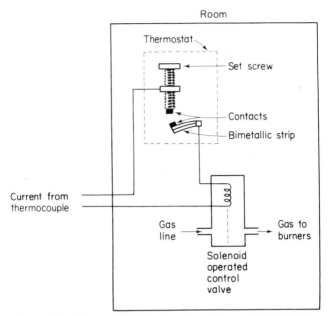

Figure 1-5 Basic discontinuous closed-loop control system.

a human operator. Even a simple closed-loop system can overcome that disadvantage. Take the example of Fig. 1-5, which is a simplified version of a typical thermostat-controlled room heater. In this case, a gas heater is used. The flow of gas to the burners is controlled by a valve which, in turn, is operated by current through a thermostat. (As a matter of interest, the current is usually generated by a *thermocouple* placed next to the heater's pilot element. Thermocouples are explained in Chapter 6. For now, we are interested in how the gas burners are controlled by a thermostat.)

As shown in Fig. 1-5, a thermostat (or controlling device) is a bimetallic strip that bends in accordance with the temperature to which it is subjected. Two different metals are bonded together. Each metal expands and contracts (with changes in temperature) at a different rate, so the strip tends to bend when temperature changes. The thermostat is mounted inside the room, usually on a wall away from a heater register. A setscrew, shown in Fig. 1-5, is adjusted so that *at the desired temperature* the contact points barely touch, and the circuit to the heater valve is completed. The desired temperature is thus the set point.

As room temperature rises above the set point, the bimetallic strip bends further from its existing position, the contact points separate, and the circuit to the heater control valve is opened. The control valve then shuts off gas to the burners, and room temperature will begin to drop. When room temperature drops below the set point, the bimetallic strip straightens

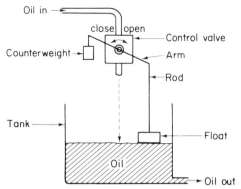

Figure 1-6 Basic continuous closed-loop control system.

enough so that the contact points touch once more. The circuit is now completed, the control valve opens, gas is applied to the burners, and the room temperature rises toward the set point. In this way, the room temperature is kept constant within a narrow *range* or *band*.

1-6 BASIC CONTINUOUS CLOSED-LOOP CONTROL SYSTEM

An example of a continuous closed-loop system is shown in Fig. 1-6. Here, the process involves keeping the oil level in a refinery tank constant, despite the fact that the tank is continuously being drained during the refining process and new oil added.

As shown in Fig. 1-6, a float attached to a rod actuates a valve through which the oil enters the tank. The set point is obtained by adjusting the float and rod so that, *at the desired level,* the valve is opened to a point where the quantity of oil entering the tank is equal to the quantity leaving the tank. When the oil level rises above a predetermined level, the float rises and reduces the valve opening so that more oil leaves the tank than enters it. When the oil falls below a desired level, the float rides lower, the valve is opened wider, and more oil enters the tank than leaves. In this way, the oil level is kept at or near the desired level.

The system shown in Fig. 1-6 is a classic example of *fully automatic control.* In the controlling device, or *controller,* the actual value of the process variable (or oil level) is compared with the value corresponding to the set point. Any difference, or deviation, is noted, as well as the phase of the difference (whether the actual value of the process variable is greater or less than the value corresponding to the set point). The controller then acts in a direction to reduce the deviation to zero.

The systems of Figs. 1-5 and 1-6 use *feedback,* or the feeding back of

7

energy from the process variable to the controlling element. In the case of the room heater (Fig. 1-5), it is the heat within the room that is applied (or fed back) to the thermostat. With the oil tank (Fig. 1-6), it is the energy of the rise or fall of the oil level that is fed back to the valve. Because of this feedback, the closed-loop system lends itself to automatic process control, thus eliminating the need for a human operator. In addition to reducing costs, this is generally more reliable since the factor of human fatigue is absent, and control is possible under conditions where a human operator could not function rapidly enough or where the environment is hazardous.

1-7 ADVANCED CLOSED-LOOP

Now let us consider a system that corresponds more closely to those found in present-day industry or other control applications. An example of such a system is shown in Fig. 1-7. Here, the system involves keeping the temperature of a process variable (a liquid in this case) at a constant temperature. The liquid flows through a container that is heated by steam. The passage of the steam to the jacket surrounding the container is controlled by a valve. In turn, the valve is controlled by the system shown in Fig. 1-7.

A transducer senses the condition of the process variable. In our system, the transducer is used to sense the temperature of the liquid and produce a corresponding electrical signal. That signal is applied to an *indicator* and to a *measuring element* in a recorder-controller. In this case, the indicator (a simple gage) is calibrated in units of the process variable, or degrees of temperature, and provides a constant readout for human operators. In more complex systems, such as shown in Fig. 1-2, the indicators are in the form of digital readouts.

A recorder-controller contains three basic elements. The first is a *measuring element,* which measures an electrical signal and converts it to some value that can be used by the recorder-controller. In some cases, the electrical signal from a transducer is converted to mechanical movement or angular rotation. In other instances, the electrical signal is converted to a numerical value or readout (possibly in pulse form) that can be used by computer circuits. The process of converting a basic electrical signal to a usable form is known as *signal conditioning.* This conditioning process can be performed in the transducer itself or in the recorder-controller or in both, depending on the system.

In many cases, the signal conditioning is a simple matter of converting from a low voltage (typically a few millivolts produced by a transducer) to some higher voltage (typically 5 volts, used by telemetry systems and computers) or conversion of direct current to alternating current and vice versa. Signal conditioning can also involve conversion from an electrical voltage (which is an analog of the process variable) to digital pulses; this is known

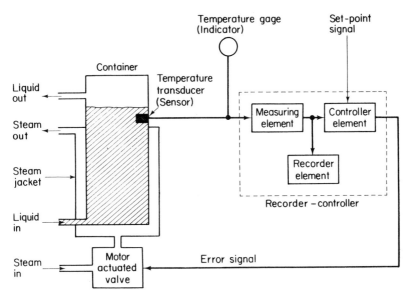

Figure 1-7 Advanced closed-loop control system.

as analog-to-digital (A/D) conversion. The reverse process is D/A, or digital-to-analog conversion. Both A/D and D/A are explained in Chapter 7.

Since it is often desirable to keep a continuous written record of a process variable condition, a typical recorder-controller contains a *recorder element* (Fig. 1-7), which continuously monitors a process variable and makes a record of it by means of a pen or moving chart. Where computers are involved, the record is usually a printout from an electric typewriter or other computer printer terminal, or possibly on magnetic tape as shown in Fig. 1-2. In simple recorder-controller systems, the signal from the measuring element positions the pen at a corresponding point on the chart. In computer-controlled systems, the signal is converted to a digital readout and then printed on the terminal.

The signal from the measuring element is also applied to the *controller element* as shown in Fig. 1-7. The controller element also receives a set point signal that corresponds to the predetermined value at which the process variable is to be kept. The controller element compares the two signals and, if there is a difference or deviation between them, determines the amount and phase of the deviation, as explained in Sec. 1-1. That is, the controller determines if the value of the process variable is greater or less than the set point value.

The controller then generates an *error signal* (generally electrical but possibly pneumatic) that corresponds to the amount and phase of the deviation. Depending on the phase of the error signal, the opening of the valve is increased or decreased, thus permitting more or less steam to flow to the container's steam jacket so as to increase or decrease the value of the process variable (the temperature of the liquid, in our case). When computers are involved, the error signal must often be converted from digital signal pulses to an electrical voltage, thus requiring D/A conversion.

No matter which system is used, the controller always acts to reduce the deviation to zero. Thus, if the actual value of the process variable rises a certain amount above the set-point value, the valve is closed and the steam is cut off. As a result, the process variable value starts to decrease. When this value falls below the set-point value, the valve is opened, the steam flows into the jacket, and the value of the process variable starts to rise again. In this way, the process variable value is kept constant between predetermined limits, or within an *error band*.

1-8 SENSING ELEMENTS OR TRANSDUCERS

From the preceding explanation it can be seen that an automatic control system starts with a primary sensing element (sensor or transducer) that senses a condition, state, or value of a process variable and produces an output which reflects a condition: typically an electrical signal output that is an analog of the process variable condition. There are many types of transducers used in present-day control systems. Among them are sensors of motion and force (acceleration, attitude, displacement, force, torque, pressure, speed, velocity, strain), fluid conditions (flow, pressure, and liquid level), humidity, moisture, light, radioactivity, temperature, and sound. Such transducers are covered in Chapters 2 through 7. Although every possible type or variation of transducer or sensor cannot be included in one book, the coverage does include a full cross section of transducers used in present-day control systems.

1-9 MEASUREMENT AND SIGNAL CONDITIONING

Measurement is an important part of any control and instrumentation system. We measure the temperature and pressure of a process variable, its rate of flow, its acidity, and much more. We need measurement not only to learn the condition or quantity of a variable, but also to obtain a value that can be compared with a standard value (the set-point value) to obtain an error signal. The error signal is used to operate an actuator, which causes the process variable to return to a predetermined value.

In a typical control system, we start by using a sensor to translate the condition or quantity of a process variable to an electrical or pressure output signal. Chatper 8 concerns some of the methods generally used to measure such signals. Typical values to be measured include current, voltage, resistance, capacitance, inductance, frequency, and pulse rate. In computer-controlled systems, the measurement process usually involves conversion from analog values to digital values and vice versa, since most of today's computers use digital pulses (although there are analog computers). A/D and D/A circuits are considered in Chapter 7.

1-10 CONTROL DEVICES

In most instances, the final stage of any control system consists of (1) a switch that may be opened or closed, (2) a valve that may be fully opened or closed or adjusted to some position between those two extremes, (3) an electromagnetic device that may be energized by an electrical current to perform some mechanical or electrical function, and (4) a motor that may be started, stopped, or reversed, or whose running speed may be varied.

Between a primary sensing element, which initiates a control system, and a final control element there may be many control elements, each performing a definite function in a system. Such devices are switches, valves, solenoids, relays, electron tubes and (in most present-day systems) solid-state control elements. Chapter 9 covers some of the most commonly used elements, explains their operation, and shows how they fit into modern control systems.

1-11 INSTRUMENTATION DEVICES

In a sense, all the devices described in this book could be called instruments or instrumentation. However, we reserve these terms to those devices that indicate, transmit, or record signals between other elements in the system, as well as the recorder-controllers. Chapter 10 concerns such instrumentation devices. In older and simpler systems, the indicators are basic pressure gages, thermometers, electrical meters, and similar devices. In more sophisticated systems, the indicators take the form of digital readouts. Therefore, in Chapter 10 we describe the basics of digital readout devices.

As mentioned, the basic recorder-controller of older and less complex systems has been replaced by digital computers in modern systems. No explanation of modern controls and instrumentation would be complete without mention of computers. However, because of the complexity of even a simple microcomputer a full description of computers is beyond the scope of this book. Here, we concentrate on control and instrumentation devices other than the computer. Emphasis is on devices that sense the process variables (the transducers), circuits that convert process variables into signals and meaningful measurements (signal conditioning and A/D circuits), circuits that convert digital commands from the computer to the control elements (D/A circuits), and to the control elements.

For a full presentation of modern computer basics, your attention is called to the author's *Handbook of Microprocessors, Microcomputers, and Minicomputers* (1979, Prentice-Hall, Inc., Englewood Cliffs, N.J. 07632).

2

MOTION AND FORCE SENSORS

Motion and force sensors can be classified by the method of sensing or by the condition they sense. In this chapter, we use the latter classification and cover devices that sense linear motion, angular motion, speed of rotation, compression, tension, torque, acceleration, vibration, and attitude. However, before going into the details of specific sensors, we describe the basics of motion and force sensing methods.

2-1 MOTION- AND FORCE-SENSING METHOD BASICS

Many control systems require sensors capable of detecting the linear motion of some moving component, the amount of movement, and its direction. Most of these sensors act by converting motion into a corresponding electrical signal and can thus be considered as *motion* transducers. The three basic types of motion transducers found in typical control systems and covered in this chapter include *linear motion, angular motion,* and *speed of rotation* transducers.

In addition to motion, many control systems must sense the presence of *force.* When a force is applied to a solid, the force may affect that solid in one or more different ways. The force may act as a *compression* tending to force molecules of the solid closer together, or the force may act as a *tension* tending to force molecules farther apart. Force may also act as a *torque* that tends to twist a solid. Thus, the basic types of force sensors covered here include compression, tension, and torque.

Acceleration can be considered as a measurement of either force or motion. When a body is in motion, if a force is applied in the direction of motion, the force tends to speed up, or *accelerate,* the body. If a force is applied in a direction opposite to the motion, the force will tend to slow down, or *decelerate,* the moving body. Acceleration sensors (or *accelerometers*) are not usually found in industrial control sytems but are used frequently for automatic control of flying vehicles such as aircraft, rockets, and missiles. Also, the principles of acceleration sensing are used to measure *vibration.*

Another force or motion used frequently for control of flying vehicles is *attitude.* The *gyro* or *gyroscope* principle is used to measure attitude or angle of a vehicle in relation to reference positions, usually horizontal, vertical, and parallel to the line of flight.

2-2 LINEAR MOTION SENSORS

The *linear-motion potentiometer,* shown in Fig. 2-1, is the most basic type of linear motion sensor and operates by converting linear motion into changes in resistance. The device is essentially a variable resistor whose resistance is varied by the movement of a slider or wiper over a resistance element. As shown, the slider is moved over the resistance element by means of an arm connected to the moving component. As the slider is moved by the arm, the resistance of the potentiometer is varied.

Typically, a slider is placed near the midpoint of a resistance element when a moving component is at its "normal" or "at rest" position. Thus,

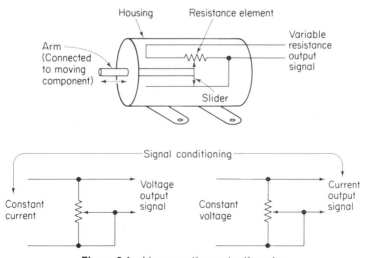

Figure 2-1 Linear-motion potentiometer.

the resistance represents the normal position of a moving component. When a component is moved to the right, less of the total resistance is shorted out, and the resistance output to the measurement or control circuits is increased. Thus, the amount of increase is a function of the amount of linear motion of a component. The opposite occurs when a component is moved to the left (the resistance decreases as a function of linear motion). The resistance remains constant when a component is stationary.

In this way, the *resistance change* is an indication of the *amount* of linear movement, whereas the *direction of movement* is indicated by whether the resistance is *increasing* or *decreasing.* The value of a sensor's output signal (resistance change) may be determined by some resistance-measuring instrument (Chapter 8) and possibly recorded. In some systems, the resistance output is applied to a controller that acts to restore a moving component to its normal position.

Not all measurement circuits, recorders, and controllers can directly accept signals in the form of resistance changes. Thus, some form of *signal conditioning* is required between a resistance element and a measurement or control circuit. If a current output signal is required, a constant-voltage source can be connected in series with a potentiometer as shown in Fig. 2-1. Then the amount and direction of movement is indicated by the change in current. If a voltage output is required, a constant-current source is used in series with a potentiometer (Fig. 2-1), and movement is indicated by the changing voltage.

The *linear-motion variable inductor,* shown in Fig. 2-2, is another basic type of linear motion sensor. Here, the inductance of the device is increased as an iron core is moved farther into the coil, and reduced as the core moves out of the coil. Since movement of the core is determined by the motion of a moving component, the inductance changes are proportional to the linear motion of the component, and the direction of these changes (increasing or decreasing) is an indication of the direction of motion.

The sensor's output signal is some value of inductance which may be determined by some inductance-measuring instrument (Chapter 8) and possibly recorded. A moving component may be controlled by using a controller capable of accepting an inductance signal.

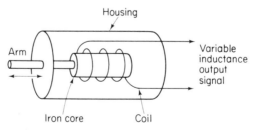

Figure 2-2 Linear-motion variable inductor.

The sensor of Fig. 2-2 can also be used as an inductor in the "tank circuit" (frequency control circuit) of an oscillator as shown in Fig. 2-3. The frequency of an oscillator is determined by a combination of inductance and capacitance. If a fixed capacitance is used, oscillator frequency then depends on the value of an inductor. Thus, an inductive value can be determined by oscillator frequency.

Linear motion can be converted to corresponding changes in capacitance by means of a *linear-motion variable capacitor,* shown in Fig. 2-4. This device consists of two plates, one fixed and the other movable, to form a capacitor using the air between them as a dielectric. The movable plate is connected by an arm to a moving component. The capacitance is increased as the moving plate is moved closer to the fixed plate and vice versa. Thus, the direction of motion is detected by an increase or decrease in capacitance. In both cases, the change in capacitance is proportional to the amount of movement.

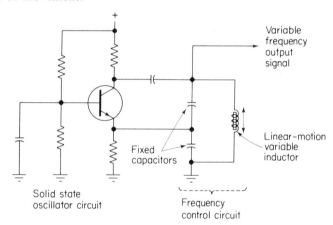

Figure 2-3 Using a linear-motion variable inductor in an oscillator frequency control circuit.

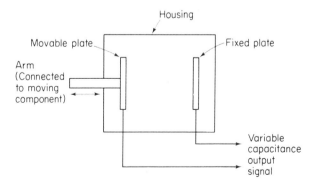

Figure 2-4 Linear-motion variable capacitor.

As with an inductive signal (Fig. 2-2), a capacitive output signal (Fig. 2-4) may be measured or recorded by means of measuring instruments and/or recorders. If a controller is used, it must be of a type capable of accepting a capacitive input signal. That capacitive signal may also be measured by using an oscillator signal in a manner similar to that shown in Fig. 2-3. However, in this case, the inductance of the oscillator is fixed, and the sensor of Fig. 2-4 furnishes the capacitance for the circuit. Thus, the capacitive value can be determined by oscillator frequency. A large change in frequency indicates a greater amount of motion. Likewise, the direction of frequency change (increasing or decreasing) indicates the direction of motion.

The *linear variable differential transformer* (LVDT), shown in Fig. 2-5, is another widely used linear-motion sensor. Here, an arm is connected to a moving component. In turn, the arm controls the position of a core within a transformer. As the core is moved back and forth by the moving component, the transformer produces an a-c output signal that reflects the *amount* and *direction* of the component movement.

Some systems require a d-c output voltage signal to match the output signals of other transducers. In that case, a signal conditioning circuit (known as a *phase-sensitive demodulator*) similar to that shown in Fig. 2-6 can be used to convert the a-c power to a d-c output. Rectifiers *CR*1 and *CR*2 are placed in each leg of the output from the transformer secondary windings. Resistor *R*1 is connected across the output from secondary *A,* and resistor *R*2 (with a resistance equal to that of *R*1) across the output from secondary *B*. The junction between *R*1 and *R*2 is connected to the junction between secondaries *A* and *B*.

When the cores is in the null or center position, the voltages across the resistors are equal and opposite, producing a net output of zero. Any physical displacement of the core causes the voltage across one resistor to increase, with a corresponding decrease in voltage across the other resistor. The *difference between the two voltages* appears across the output terminals and produces a measure of the physical position of the core (and the moving

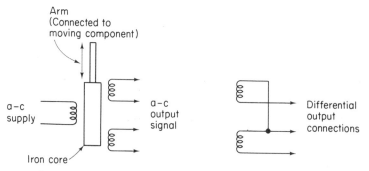

Figure 2-5   Linear variable differential transformer (LVDT).

component). The *amplitude* of the d-c voltage indicates the *amount of deviation* from the null or center position. The *polarity* of the d-c output indicates the *direction of movement* from the null position.

2-3 ANGULAR MOTION SENSORS

The *angular-motion potentiometer,* shown in Fig. 2-7, is the most basic of the angular motion sensors. The potentiometer of Fig. 2-7 resembles the linear-motion potentiometer shown in Fig. 2-1, except that the resistance element is circular instead of straight. The slider is mounted on a shaft, and as the shaft is rotated the slider moves over the resistance element.

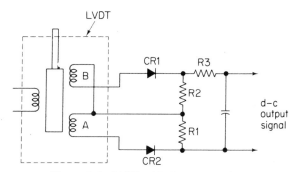

Figure 2-6   LVDT with d-c output signal.

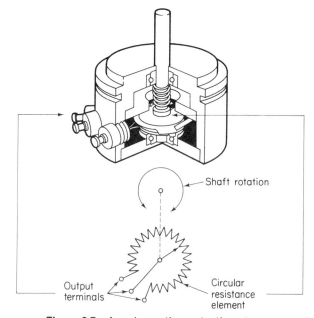

Figure 2-7   Angular-motion potentiometer.

The rotating component is coupled to this shaft and, as the component rotates, the resistance of the potentiometer changes in proportion to the angular motion. The direction of rotation can be determined by whether the resistance is increasing or decreasing. Typically, the angular motion potentiometer is suitable only for angular motion not exceeding about 300°.

The *angular-motion variable capacitor,* shown in Fig. 2-8, is another basic angular motion sensor. The capacitor of Fig. 2-8 consists of a metal plate that moves between two fixed metal plates as a shaft is rotated. The three plates, and the air between them, form a capacitor with a capacitance that varies in proportion to the degree to which the plates are meshed. When the plates are completely meshed, the capacitance is at its maximum. When the plates are completely unmeshed, the capacitance is at minimum.

The rotating component is coupled to the capacitor shaft. Thus, the capacitance is a function of the degree of angular rotation. Also, the direction of rotation may be determined by whether the capacitance is increasing or decreasing. Typically, the angular motion capacitor is suitable only for angular motion not exceeding about 300°.

The *rotary variable differential transformer* (RVDT), shown in Fig. 2-9, is a variation of the linear variable differential transformer shown in Fig. 2-5. The RVDT of Fig. 2-9 is similar to the linear type, except that the core of the RVDT is cam-shaped and is rotated between the transformer windings by means of a shaft. At the null position of the core, the voltage outputs from the two secondaries are equal but opposite in phase, producing a net output of zero volts. Any rotary displacement from this null position results in a differential-voltage output. The greater the rotary displacement, the larger the differential-voltage output.

Clockwise rotation of the shaft produces an increasing voltage of one phase. Counterclockwise rotation also produces an increasing voltage, but of the opposite phase. Thus, both the amount of angular motion and the

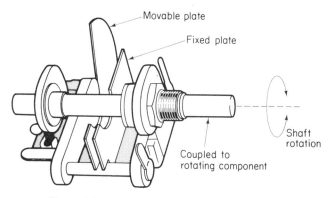

Figure 2-8 Angular-motion variable capacitor.

direction can be obtained from the voltage and phase of the RVDT output signal. For those systems that require a d-c output voltage signal to match the output signals of other transducers, the signal conditioning circuit shown in Fig. 2-9 is used. Operation of this circuit is similar to that of the LVDT of Fig. 2-6.

The *variable-reluctance transducer,* shown in Fig. 2-10, is another angular-motion sensor somewhat similar to the RVDT. The basic elements of this transducer are two identical coils (L1 and L2), connected in series op-position, and a cam-shaped soft-iron armature. The coils and armature are so positioned that when the armature is in the null or zero position, the im-pedances of both coils are the same. The voltage drops across L1 and L2 are equal but of opposite phase in the zero position.

Any rotary displacement of the shaft and armature from the null posi-tion results in a differential-voltage output. The amplitude of this voltage is directly proportional to the amount of displacement. The phase of the out-put voltage indicates the direction of the angular displacement. As in the case of the LVDT and RVDT, a signal conditioning circuit is used in systems that require a d-c output voltage.

Typically, the RVDT of Fig. 2-9 and variable-reluctance transducer of Fig. 2-10 are suitable for measurement of angular motion not exceeding about $\pm 45°$.

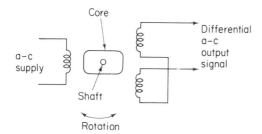

Figure 2-9 Rotary variable differential transformer (RVDT).

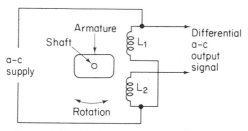

Figure 2-10 Variable-reluctance transducer.

2-4 SPEED OF ROTATION SENSORS

There are a number of methods for sensing the speed of rotation of some rotating component such as a motor. One of the most basic methods is the electrical *tachometer* shown in Fig. 2-11. With such a tachometer, the shaft of the rotating component is coupled to the shaft of a small d-c generator. The output voltage of a generator is fed to a voltmeter. The scale of the voltmeter is calibrated in units of rpm (revolutions per minute). The amplitude of generator output voltage is directly proportional to the rotational speed of the rotating component. Thus, an increase in rotational speed produces a high voltage which, in turn, appears as a higher rpm indication.

Some tachometers use an a-c generator. In such systems, the frequency of the voltage output is a function of the speed of rotation. Such systems are more complex since they require that the indicator be capable of measuring increases and decreases in frequency, rather than simple changes in d-c voltage, as is the case with the d-c generator tachometer.

The *rotating disk and light sensor,* shown in Fig. 2-12, is another method for measuring speed of rotation. The basic elements of this system

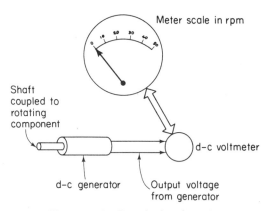

Figure 2-11 Electrical tachometer.

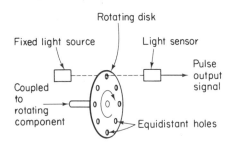

Figure 2-12 Speed of rotation sensor using rotating disk and light sensor.

are a perforated rotating disk coupled to the rotating component by a shaft, a light source, and a light sensor. As shown in Fig. 2-12, the disk has a number of equidistant holes or perforations. A fixed light source is placed on one side of the disk, in line with the holes. A light sensor (such as a photo transistor described in Chapter 5) is placed on the opposite side of the disk but also in line with the holes and light source. When the solid or opaque portion of the disk is between the light source and the sensor, no light is applied to the sensor, and no output voltage is produced. When the disk is rotated so that a hole appears between the sensor and light source, the sensor produces an output voltage. If the disk is rotating steadily, the sensor produces "pulses" of voltage (one pulse for each hole). These pulses are similar to the voltage pulses found in computers and other digital electronic equipment.

The frequency at which the pulses are produced depends on the number of holes in the disk and its speed of rotation. Since the number of holes is fixed, the pulse rate thus is a function of rotational speed. The pulses can be converted to an rpm indication by a frequency measuring device or counter such as described in Chapter 8. For example, if a light sensor is connected to a frequency counter and produces an indication or count of 300 pulses per minute, the rotational speed is equal to 300 divided by the number of holes. Assuming that there are six holes in the disk, the rotational speed is 50 rpm (300 pulses per minute divided by 6).

For those systems that require a d-c output voltage signal to match the output signals of other transducers, a signal conditioning circuit is used between the light sensor and the system. The digital-to-analog (D/A) converter described in Chapter 7 is such a conditioning circuit. The D/A circuit converts voltage pulses from a light sensor into a d-c voltage that is proportional to the rpm.

The *magnetic pickup,* shown in Fig. 2-13, is a similar and possibly more common method for measuring speed of rotation. The basic elements of this system are a metallic toothed wheel, such as a gear wheel, connected by

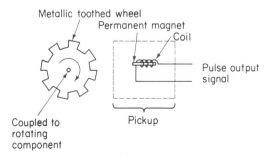

Figure 2-13 Speed of rotation sensor using magnetic pickup.

a shaft to the rotating component, and a magnetic pickup or coil. As shown, the wheel teeth pass near the pickup coil. Typically, the pickup consists of a housing containing a small permanent magnet with a coil wound around it. A fixed magnetic field surrounds the pickup. Since both the field and coil are stationary, no voltage is induced in the coil. When the wheel is rotated and the teeth pass through the field, a voltage pulse is induced in the coil.

The frequency of the pulses depends on the number of teeth and the speed of rotation. Since the number of teeth is known, the pulse frequency can be related directly to rpm. The rotational speed is equal to the pulse frequency divided by the number of teeth. As in the case of the rotating disk, the pulses can be converted to a d-c voltage by means of a D/A circuit, or the pulses can be measured by a frequency counter.

2-5   COMPRESSION SENSORS

One of the most common compression or force sensors is the *bonded wire strain gage* shown in Fig. 2-14. This device consists of a length of fine wire arranged in the form of a grid and bonded to a paper or plastic sheet.

The resistance of a wire depends on its composition, temperature, length, and cross-sectional area. If all other factors are kept constant, the greater the length of the wire the greater its resistance; the shorter its length, the smaller its resistance. Conversely, the smaller the cross-sectional area of the wire, the greater its resistance; and the greater its area, the smaller its resistance.

This change of resistance is the principle upon which a strain gage operates. A typical strain gage application which illustrates the operating principle is that of the *load cell* shown in Fig. 2-15. In this system, the gage is cemented to the side of a steel column. The column is enclosed in a metal housing so that only the end of the column protrudes through the top. Electrical connections to the ends of the gage are made through the side of the housing.

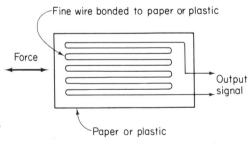

Figure 2-14   Bonded wire strain gage.

When heavy pressure is applied at the top, the column and the strain gage cemented to the side are compressed. As the strain gage is compressed, the length of its wire is shortened and its cross-sectional area is increased, producing a decrease in strain gage wire resistance. The decrease is directly proportional to the pressure on the top of the load cell. By measuring the change in resistance we can determine the amount of pressure.

Note that the strain gage is acting as a *secondary transducer*. Actually, the steel column is the primary sensing element because it translates the variable (or pressure) into variations of its physical dimensions. The strain gage translates these changes in physical dimensions into changes in resistance.

Load cells are generally used to measure relatively large forces. For example, the force acting on an upright beam supporting a bridge or some other structure may be measured with a load cell. Another typical application for a load cell is to measure the weight of a liquid in a tank by measuring the force the liquid exerts on the tank.

By itself, a strain gage produces a resistance change that is proportional to force or compression. Most systems require a voltage output (proportional to force) that is compatible with other transducers. This can be accomplished by means of a *bridge-type signal conditioning* circuit such as shown in Fig. 2-16. (Bridges are discussed further in Chapter 8). The strain gage resistance forms one arm of the bridge in Fig. 2-16. The power supply may be either a-c or d-c voltage.

In a typical application, the bridge is balanced by adjustment of R1 so that there is no output voltage when there is no force applied to the strain gage. When force is applied, the resistance of the strain gage changes, and

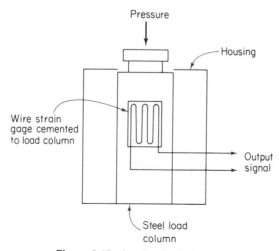

Figure 2-15   Load cell strain gage.

the bridge becomes unbalanced, producing an output voltage. The magnitude of that voltage depends on the resistance change of the gage which, in turn, depends on the force applied to the gage. In some applications, the output voltage is measured by a voltmeter with a scale calibrated in terms of force (grams, pounds, ounces, etc.).

One problem with a wire strain gage is that temperature changes will also affect its resistance. To compensate for such temperature changes, two similar gages (RA and RB) can be mounted on a load cell or other device as shown in Fig. 2-17. One gage is mounted in line with the force to be measured, and the other gage is mounted at right angles to the force. Both RA and RB are subject to the same temperature and undergo similar resistance changes as a result. Since RA and RB are in opposite arms of the

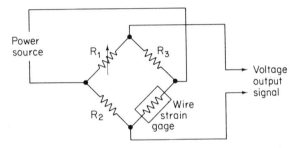

Figure 2-16 Basic bridge-type signal conditioning circuit.

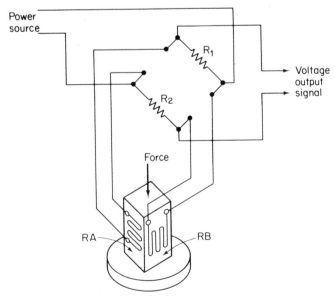

Figure 2-17 Compensated strain-gage bridge circuit.

bridge, the resistance change due to temperature of one balances out the change in the other. However, because one gage is at right angles to the other (and to the force) the resistance change due to force will be different for each gage, and this difference will produce an output voltage (proportional to the force).

Unbonded wire strain gages are often used to measure small amounts of force. A typical arrangement is shown in Fig. 2-18. That strain gage consists of a cross-shaped spring, firmly fixed at its four ends. Posts are set in each arm of the spring and two wire windings are wound around the posts, one winding over the spring and the other below. A force rod transmits the force to the center of the spring.

When a force is applied through the rod to the spring, the movement of the spring causes a corresponding movement of the posts. As a result, the strain (and the resistance) of the winding is increased. Simultanously, the strain and resistance of the other winding is decreased. The two windings form two arms of a normally balanced bridge as shown in Fig. 2-18. When a force is applied and the resistances change, the bridge becomes unbalanced, producing an output voltage that is proportional to the force.

Semiconductor strain gages are another form of force sensor. There are two basic types of semiconductor force sensors: the *piezoresistance* sensor

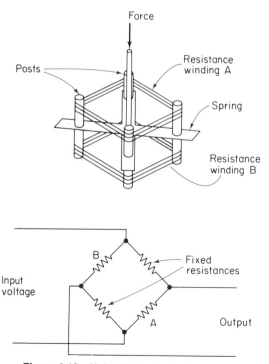

Figure 2-18   Unbonded wire strain gage.

and the *piezoelectric* sensor. Both types operate on the piezo-effect of semiconductor materials.

For the piezoresistance sensor shown in Fig. 2-19, a semiconductor material is placed between two plates. A typical material is silicon crystal similar to that used in silicon transistors. When force is applied to such material, by pressing the plates together the resistance of the material changes in proportion to the amount of force. In effect, the plates and silicon material become a variable resistance and can be connected in a typical bridge, as described for the wire strain gage. The applied force produces a proportional output voltage.

For the piezoelectric sensor shown in Fig. 2-20, another kind of semiconductor material is placed between two plates. Typical materials are quartz crystals, Rochelle salt, and barium titanate. When force is applied to such material, by pressing the plates together, a current is generated through the material and plates. In effect, a voltage (proportional to force) is developed across the plates. That voltage can be used directly. One way, for example, would be to apply the voltage to a voltmeter having a scale marked in units of force (grams, ounces, etc.). However, since the voltage developed is very small, it is usually more practical to amplify the voltage through signal conditioning circuits to a level compatible with other transducers in the system.

One problem with the piezo force sensors, particularly the piezo-resistance sensor, is that they are quite vulnerable to temperature changes.

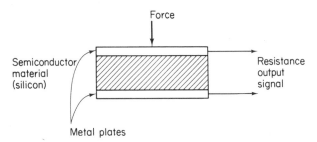

Figure 2-19   Piezoresistance (semiconductor) strain gage.

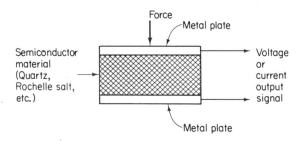

Figure 2-20   Piezoelectric (semiconductor) strain gage.

For that reason, piezoresistance sensors are often used in pairs, as described for the wire strain gage. The main advantage of the piezo force sensor is its extreme sensitivity; that is, the piezo sensor is able to produce a usable output signal (voltage or resistance) with a very small force. Also, the characteristics of the piezoresistance sensor material (silicon) can be altered to meet specific sensitivity requirements by varying the amount of impurities in the silicon chip (a process known as "doping"), just as is done in silicon transistors.

2-6 TENSION SENSORS

Bonded and unbonded wire strain gages can be used as tension sensors. The crane scale shown in Fig. 2-21 is an example of how a bonded strain gage can be used to measure the tension produced by a given load on a crane. In this case, the strain produced by the heavy load stretches a metal column to which the strain gage is cemented. As a result of the stretching, the wire becomes longer and its cross-sectional area smaller, thus producing an in-

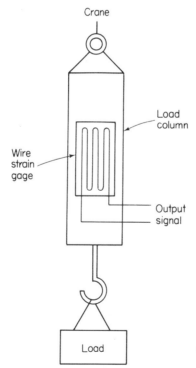

Figure 2-21   Tension sensor (crane scale).

crease in resistance. Since the increase is directly proportional to the strain producing it, the strain is found by measuring the change in resistance. An unbonded strain gage can be used in a similar manner and is particularly effective for measuring small amounts of tension.

Semiconductor strain gages, both piezoresistance and piezoelectric, can also be used as tension sensors.

2.7 TORQUE SENSORS

As rotary motion is applied to a shaft, a twisting force called *torque* is imparted to the shaft. A bonded wire strain gage can be used as a sensor of that force as shown in Fig. 2-22. This sensor consists of a short shaft called the *torque shaft,* one end of which is attached to the shaft under test and the other end to the device that is producing the rotation. A bonded wire strain gage is cemented to the side of the torque shaft and, as the shaft is rotated, the twisting action of the torque stretches the wire of the gage, increasing its resistance. Since the increase in resistance is directly proportional to the torque producing the increase, the torque or force can be found by measuring the change in resistance.

In the sensor system of Fig. 2-22, the ends of the strain gage are brought out to the external circuit by means of slip rings and brushes similar to those used on an a-c motor or generator. As a result, the shaft may rotate without snapping the connecting leads.

2-8 ACCELERATION AND VIBRATION SENSORS

Due to inertia, if a body or mass is at rest, the body tends to remain at rest. If the body is in motion, the body tends to maintain that motion in a straight line. If force is applied to the body, the force tends to overcome inertia. If the body is at rest, the force will move the body in the direction that the force is applied. When the body is in motion, if the force is applied in the direction of motion, the body tends to speed up, or *accelerate.* If the force is applied in a direction opposite to the motion, the body tends to slow down, or *decelerate.*

Where automatic control is required of vehicles such as rockets, unmanned aircraft, or space probes, there must be some method for sensing and measuring their force of acceleration. This generally is done by means of acceleration sensors, or *accelerometers.* There are many types of accelerometers: one of the most common uses the piezoelectric effect of quartz crystals described in Sec. 2-5. A typical piezoelectric crystal type of accelerometer used in aircraft is shown in Fig. 2-23. In this arrangement, the crystal is placed between two metal electrodes, the crystal-electrode package is fastened to a base (which could be part of the aircraft), and a fixed mass or body is placed on top of the package.

The mass exerts a certain force on the crystal and, as a result of that force, a certain voltage is generated. As the base of the accelerometer is accelerated in the direction indicated, the force exerted by the mass on the crystal is increased, due to the inertia of the mass, which tends to oppose the acceleration. Acceleration is equal to force divided by mass. Since the mass is a fixed quantity, the increase of the force that the mass exerts on the crystal is proportional to the acceleration. Thus, the output voltage is proportional to acceleration.

The change or increase in the voltage output from the crystal may be measured by a voltmeter. However, the usual arrangement in aircraft or

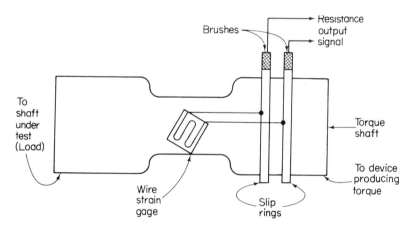

Figure 2-22 Torque sensor.

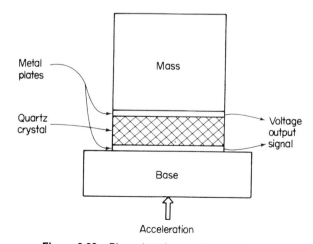

Figure 2-23 Piezoelectric accelerometer.

space applications is to use the output as an analog of acceleration and apply the output to a computer that controls operation of the aircraft. Acceleration generally is measured in units of *g*'s (*gravity*). If a mass weighing 150 pounds is placed on the floor of a stationary aircraft, the mass will press down on the floor with a force of 150 pounds, due to the gravitational attraction of the earth. Should the aircraft then be accelerated away from the earth so that the force with which the mass presses down on the aircraft's floor is 300 pounds, the acceleration increases to 2 *g*'s. At 450 pounds, the acceleration is 3 *g*'s, etc. The accelerometer's output voltage may be related in units of *g*'s. The *g*-units indicate how many times greater the force of the mass is upon the accelerometer when the aircraft is accelerated than when the aircraft is at rest.

Not all accelerometers are used to measure the acceleration of space vehicles. Accelerometers can also be used effectively to monitor *vibration*. When a moving body is subject to vibration that body is, in effect, accelerated alternately in one direction and then the other. Thus, a basic accelerometer can be used affectively as a *vibration sensor*.

The *inductive vibration sensor* shown in Fig. 2-24 is a typical vibration sensor. One end of a flexible reed is fastened to the side of the housing, and a mass is fastened to the free end of the reed as shown. A coil of wire, wound upon a nonmagnetic form, is attached to the base of the housing. A bar magnet is connected to the reed by means of a rod so that the magnet may move in or out of the coil as the reed vibrates. The base of the housing is attached to the device whose vibrations are to be measured.

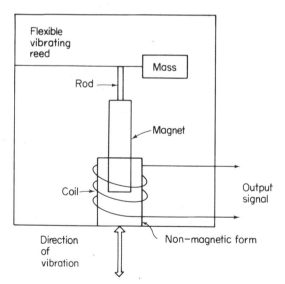

Figure 2-24 Inductive vibration sensor.

As the device vibrates, the entire sensor is moved up and down in the directions indicated by Fig. 2-24. When both the coil and magnet are stationary with respect to each other (no vibration), there is no voltage induced in the coil. Vibration produces relative motion between the coil and magnet. In turn, the field around the magnet cuts across the winding, and a voltage is induced in the coil. The greater the relative motion, caused by vibrations with greater amplitudes, the greater the induced voltage. Thus, the magnitude of the sensor's output voltage is a measure of the vibration amplitude.

The *capacitive vibration sensor* shown in Fig. 2-25 acts in a manner similar to that of the inductive vibration sensor. With the capacitive sensor, vibrations cause variations in capacitance of a variable capacitor. The capacitor consists of a movable plate attached to the mass, and a fixed plate attached to the base of the sensor.

The output of the basic sensor is a variable capacitance. The amplitude of vibration is measured by the amplitude of the capacitance variations (how far the capacitance varies from a fixed, no-vibration value). Elaborate signal conditioning circuits are usually required to convert capacitance variations into a voltage indication that is compatible with other sensors. For that reason, a capacitive vibration sensor is seldom used except for special cases. One method for measuring the amplitude of vibration is to use the capacitor to frequency-modulate on oscillator. The demodulated output of the oscillator then furnishes an indication of the amplitude of vibration.

A more common type of vibration sensor uses a linear variable differential transformer of the type discussed in Sec. 2-1.1. Such a vibration sensor is shown in Fig. 2-26. Here, the core of the transformer is maintained in its null position by means of rods connected to flexible reeds at either end. The core itself is the mass of this sensor. As the sensor moves up and down as a result of vibration, the output voltage from the transformer is an a-c signal.

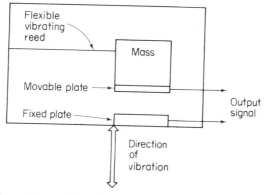

Figure 2-25 Capacitive vibration sensor.

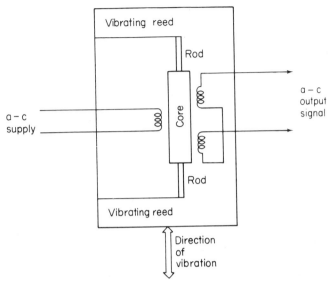

Figure 2-26 LVDT vibration sensor.

The phase of the signal shifts alternately in one direction and then the other, in response to each vibration. The magnitude of the output signal depends on the amplitude of the vibration.

As in the case of other sensors using a variable differential transformer, the output signal of a vibration sensor may be rectified to produce a d-c voltage that is compatible with other sensors. The magnitude of the d-c voltage is an indication of vibration amplitude.

2-9 ACCELERATION SENSORS (ACCELEROMETERS)

There are several types of accelerometers in addition to the basic systems described in Sec. 2-8. Before going into the details of specific accelerometer types, let us consider the units of measurement and the basic terms common to all types.

2-9.1 *Units of Acceleration Measurement*

The unit of *linear acceleration* is g, or the acceleration produced by the force of gravity. Since the earth's gravity varies with the point of observation, the value of g has been standardized by international agreement as follows:

$$1 \text{ (standard) } g = 386.087 \text{ in./s}^2 = 32.1739 \text{ ft/s}^2 = 980.665 \text{ cm/s}^2$$

From this, it can be seen that acceleration is measured by an amount of motion divided by a period of time. Generally, g is simplified to

$$386 \text{ in.}/s^2 = 32.2 \text{ ft}/s^2 = 981 \text{ cm}/s^2$$

Shock and vibration amplitudes are also given in terms of g.

Angular acceleration (or rotational acceleration) is measured in rad/s^2, or radians per second squared. One degree is equal to 0.01745 radians. One radian equals 57.296° (rounded off to 57.3°).

Power density, which is a term used primarily for vibration measurement as discussed in Sec. 12-8, is usually expressed in g^2/Hz, where Hz is hertz.

2-9.2 Basic Acceleration Measurement Terms

Acceleration is a time rate of change of velocity with respect to a reference point. Acceleration is usually measured by means of a *seismic system,* consisting of a mass suspended from a reference base by a spring. *Damping* in a seismic system is an energy-dissipating characteristic which tends to bring a system to rest when a stimulus is removed. *Shock* (which usually refers to mechanical shock) is a sudden nonperiodic or transient excitation of a mechanical system. *Jerk* is a time rate of change of acceleration with respect to a reference point.

Vibration is a form of acceleration. Mechanical vibration is an oscillation where the quantity is mechanical in nature (such as force, stress, velocity, acceleration, displacement). *Harmonic* vibration or motion is a vibration whose instantaneous amplitude varies sinusoidally with time. *Random* vibration is nonperiodic vibration. *Periodic* vibration is vibration which repeats itself at certain equal increments of time. A *period* is the smallest increment of time for which the waveform of a periodic vibration repeats itself. A *cycle* is the complete sequence of a periodic vibration that occurs during one period. The *frequency* of a periodic vibration is the reciprocal of the period. The *phase* of a periodic vibration is the functional part of a period through which the vibration has advanced, as measured from a reference point. The *frequency spectrum* of vibration is a description of the instantaneous content of components, each of different frequency and usually of different amplitude and phase. *Power density* of random vibration is the mean square magnitude per unit bandwidth of the output. Power density is usually considered as the output of a unity-gain filter responding to a vibration. A *power density spectrum* is a graphical representation of values of power density displayed as a function of frequency so as to represent the distribution of vibration energy with frequency.

2-9.3 Basic Acceleration-Sensing Elements

The basic sensing element of most accelerometers is a *seismic mass,* also known as a *proof mass.* Usually, the mass is restrained by a spring and a damping system as shown in Fig. 2-27. When acceleration is applied to the accelerometer, the mass moves relative to the case as shown. When acceleration stops, the spring returns the mass to its original position.

The seismic mass of a linear accelerometer is typically a circular or rectangular body, arranged to slide along a bar and restrained from motion in all directions except the sensing axis. The mass of an angular accelerometer is typically a disk, pivoted at its center and restrained by a spiral spring so as to measure angular acceleration (acceleration with an angular displacement). Note the black/white circular symbol shown on the mass in Fig. 2-27. That symbol is commonly used to point out the location of the seismic mass gravitational center.

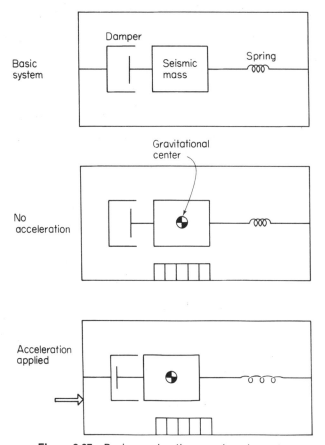

Figure 2-27 Basic acceleration sensing elements.

If acceleration is constant, the displacement of the mass can be expressed by the simple equation $y = aM/k$, where y = displacement in cm, a = acceleration in cm/s^2, M = mass in grams, and k = the spring constant in dynes/cm. If acceleration is varying, the damping constant k must be modified.

2-9.4 Piezoelectric Accelerometers

Piezoelectric accelerometers operate on the principle that certain crystal materials produce an electrical charge or voltage when compressed or squeezed between conductors. The crystals can be natural or synthetic. *Quartz* is the most commonly used natural crystal for accelerometers and other piezoelectric transducers. Many kinds of synthetic crystals are used. Usually, the process for manufacturing synthetic crystals is kept secret, although virtually all such crystals are a form of ceramic. An important characteristic of piezoelectric ceramic crystals is the *Curie point,* or the temperature at which the crystal loses its ability to produce an electrical charge (loses it polarization) when heated. Thus, the Curie point sets the upper limit of operating temperature for the accelerometer or transducer. Some piezoelectric ceramics can be repolarized after they are heated beyond the Curie point, but the transducer will still stop operating when the crystal is reheated beyond the upper temperature limit. Typical limits are 120 °C for barium titanate and 570 °C for lead metaniobate.

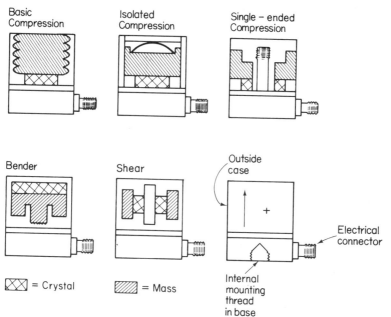

Figure 2-28 Typical piezoelectric accelerometer designs.

Typical piezoelectric accelerometers are shown in Fig. 2-28. These designs are commonly used for vibration sensors but can also be used in other linear applications. Note the arrow and " + " symbol on the outside of the case. The arrow indicates the *sensing axis* of the accelerometer (the direction in which acceleration is measured). The " + " symbol indicates the direction in which the case must be accelerated to get an output of positive polarity. In all of the Fig. 2-28 designs, the sensing axis is at right angles to the base, and the crystals are polarized so as to minimize any output due to acceleration along all other axes.

In all of the designs of Fig. 2-28, a part of the spring action is provided by the crystal itself. In the *basic* and *isolated-compression* designs, the case also contributes to the spring or elastic function. The isolated-compression design includes a curved spring which adds to the spring action. Since the case is part of the spring function, operation of both the basic and isolated-compression designs is affected by any force acting on the case. This condition is known as *case sensitivity*.

Case sensitivity can be minimized (but not eliminated) by using one of the other designs shown in Fig. 2-28. The *shear design* uses an annular crystal bonded to a center post on its inside surface and to an annular mass on its outside surface. Up and down deflection of the mass causes shear stresses across the crystal. In the *compression design,* the mass is preloaded against the crystal by a spring so that an output of alternating polarity is produced when compression is alternated by vibration. In the *bender design,* the crystal is bonded to a mass that operates as an elastic member. The rim of the assembly deflects up and down with respect to the supported center.

In a typical piezoelectric accelerometer, the base and case are electrically connected to one crystal electrode, and the case is hermetically sealed to keep out moisture. Any moisture on the crystal can produce electrical leakage and reduce accelerometer output. As shown in Fig. 2-28, the case is provided with an electrical connector to complete connections between the crystal and output circuit. As shown, the most common form of mounting is an internal thread in the base. A special insulated mounting stud can be used if the case must be electrically insulated from the mounting surface or structure. Very small cases are cemented to the structure.

Signal conditioning.   Piezoelectric accelerometers almost always require special signal conditioning circuits because the output voltage of a crystal is small but the impedance is high, in relation to that found in accelerometers using a resistance or potentiometer as the sensing element. The low voltage can be increased by means of an amplifier. In some cases the amplifier can be mounted within the accelerometer case, particularly if an IC (integrated circuit) amplifier is used. The high impedance problem can be overcome by means of an emitter-follower. A typical piezoelectric signal conditioning circuit is shown in Fig. 2-29. With such a circuit, the impedance is reduced to the same level as any other transistor amplifier, and the output voltage is increased many times.

In those accelerometers where the crystal is connected directly to the electrical connector, without an amplifier or impedance-reducing circuit, the cable can present a signal conditioning problem. Since the cable carries the high-impedance output of the crystal, it must be as free as possible from *triboelectric noise* (which is noise induced by friction between conducting and insulating portions of the cable). To keep cable noise at a minimum, the cable should be flexible, thin, coaxial, shielded, moisture-resistant, and preferably of low capacitance. It should be noted that the most common failure of piezoelectric accelerometers is breakage of the cable.

2-9.5 Resistance Accelerometers

In this kind of accelerometer, a seismic mass is mechanically linked to a wiper arm of a variable resistor or potentiometer. The arm moves over the active portion of the resistance element with full-span deflection of the mass. A typical resistance accelerometer design is shown in Fig. 2-30. Damping is generally used to minimize vibration-induced noise in the output due to wiper whipping and large changes in instantaneous contact resistance between wiper and resistance element. The damping can be magnetic, viscous (oil-filled), or gas. With magnetic damping, the mass is operated within the field of a permanent magnet. With viscous or gas damping, the mass is attached to an oil or gas-filled bellows.

Most resistance accelerometers are of the bidirectional-range design. That is, the mass can move in both directions along the sensing axis, and the

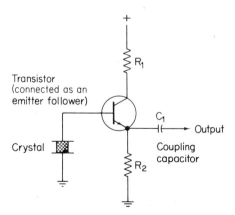

Figure 2-29 Typical piezoelectric accelerometer signal conditioning circuit.

wiper is at the center of the resistance element if no acceleration is applied. Typically, there are overload stops to limit wiper travel if accelerations exceed the accelerometer range. The total stroke of the wiper contact over the resistance element is quite small. For that reason, the wiper may be mechanically linked in such a way that a large change in acceleration produces a small change in wiper movement.

2-9.6   *Reluctance Accelerometers*

The variable-reluctance principle discussed in Sec. 2-3 can also be used for accelerometers. There are two basic types of reluctance accelerometers: the *inductance bridge* type, and the *differential transformer* type (which is described in Sec. 2-8 and shown in Fig. 2-26).

A typical reluctance accelerometer of the inductance bridge type is shown in Fig. 2-31. With this system, two coils are mounted over a laminated core to which a mass is attached. When acceleration acts on the mass, the core deflects or moves so that the inductance of one coil increases while that of the other coil decreases. The two coils are connected as a half-bridge, with the other half formed by two resistors of equal value. The resistors can be made variable, so the output of the bridge is zero when the core is centered (no acceleration).

When an a-c voltage is connected across the bridge, an a-c output is produced by unbalance of the bridge (if acceleration moves the core away from

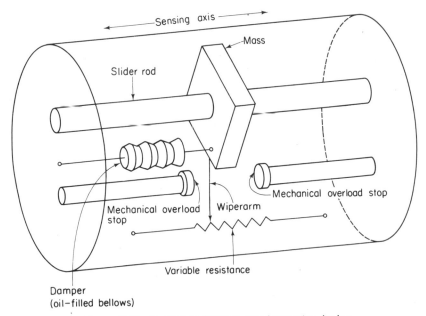

Figure 2-30   Typical resistance accelerometer design.

center). Equal upward and downward motions of the mass (typical of vibration) produced output changes of equal amplitude but of opposite phase. A phase-sensitive demodulator can convert this output into a d-c voltage whose polarity indicates the direction of acceleration. Such phase-sensitive demodulators are discussed further in Sec. 2-2.

2-9.7 Strain-Gage Accelerometers

The strain-gage principle discussed in Sec. 2-5 can also be used for accelerometers. For example, the unbonded strain gage illustrated in Fig. 2-18 can be converted to an accelerometer if the force rod is connected to a suitable mass. When acceleration occurs, the mass applies force through the rod to the spring. Movement of the spring causes a corresponding movement of the posts, thus increasing the strain and resistance of one winding and decreasing the resistance of the other. Typically, the two windings form two arms of a normally balanced bridge as shown in Fig. 2-18. When acceleration is applied and the resistances change, the bridge becomes unbalanced, producing an output voltage that is proportional to the acceleration.

In some strain-gage accelerometers, the wires themselves supply spring force in the spring-mass system. Such a system is shown in Fig. 2-32. Here,

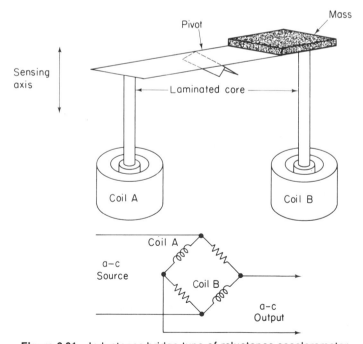

Figure 2-31 Inductance-bridge type of reluctance accelerometer.

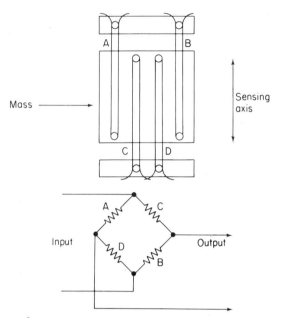

Figure 2-32 Strain-gage type of accelerometer.

four strain-gage windings are wound between insulating studs on the mass and on the frame of the accelerometer case. The rectangular mass is mechanically restrained from motion in all directions except the sensing axis. As the mass is moved in one direction by acceleration, one set of two windings is stretched while the opposite set of windings is compressed, thus increasing the resistance in one set of windings and decreasing the resistance in the opposite set. Again, the windings are connected in some form of bridge circuit. The amplitude of the bridge circuit output represents the amount of acceleration, whereas the polarity of the output voltage represents the direction of acceleration.

2-10 ATTITUDE SENSORS

Attitude sensors are used primarily in aircraft, missiles, rockets and other space vehicles. Generally, the attitude sensor is part of the control or guidance system but can also be used for measurement during flight. Attitude sensors are also used as a gravity-sensing transducer in measurement systems where small deflections from true horizontal or vertical are to be monitored. The *autopilot* found in many aircraft applications uses various forms of attitude sensors (often gyro-based attitude sensors), as does the *inertial-guidance* system of space vehicles. Before going into the details of the various kinds of specific attitude sensors, let us consider the units of measurement and the basic terms common to all types.

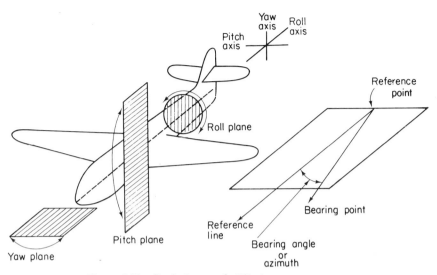

Figure 2-33 Basic terms of attitude measurement.

2-10.1 Basic Terms and Units of Attitude Measurement

Attitude is the relative orientation of a vehicle (or any object), represented by its angles of inclination to three reference axes. As shown in Fig. 2-33, an aircraft is represented as having three angles of inclination: *pitch, yaw,* and *roll.* The attitude of the aircraft can then be represented as pitch, yaw, and roll attitude. Yaw is sometimes known as *sideslip.*

The units of measurement for attitude are *angular degrees, minutes, and seconds.* For some cases, *radians* (Sec. 2-9.1) are used for attitude measurement.

Attitude sensors (often called *attitude transducers*) measure the angles of pitch, yaw, and roll. *Attitude-rate* transducers or sensors measure the *time rate of change* of attitude, often referred to as "rate of rotation" or, more simply, "rate". The units of measurement for attitude-rate are usually *degrees per second.*

The terms *bearing* and *azimuth* represent directions from a given reference point, as shown in Fig. 2-33. Usually, bearing is given by an angle, in the horizontal plane, between a reference line and the line between the reference point and the point whose bearing is specified.

2-10.2 Attitude-Sensing Methods

There are five basic attitude-sensing methods: *gravity reference, magnetic reference, flow-stream reference, optical reference,* and *inertial reference* (also known as *gyro reference.)* Gravity-, magnetic-, flow-stream, and inertial-reference attitude sensors are often used on aircraft and missiles

41

operating within the earth's atmosphere. When the aircraft is operated outside the earth's atmosphere and away from the earth's gravitational pull and magnetic field, an inertial- or gyro-reference attitude sensor is used alone. An optical-reference attitude sensor is a highly specialized transducer (usually a transducer and related control system) designed for specific outer space missions. Because of their highly specialized nature, we do not discuss optical-reference attitude sensors here.

2-10.3 Gravity-Reference Attitude Sensors

A vertical reference point (to measure roll and pitch) can be established by using the force of gravity acting on a mass. One of the simplest and oldest gravity-reference sensors is a plumb bob attached to a string. In construction, a plumb bob and string are used to check whether a structure is vertical, or exactly straight up and down. In use, a plumb bob is suspended by a string and held next to a structure (say a wall stud). The structure is exactly parallel to the string if the structure is truly vertical.

This same principle can be used in the pendulum-resistance attitude sensor shown in Fig. 2-34. In this example, the pendulum consists of a weight pivoted to rotate freely about the center of the sensor case in a ball bearing. Rotation of the pendulum causes a wiper to travel over a resistance element.

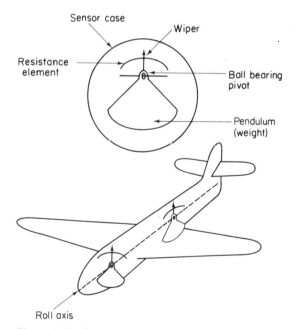

Figure 2-34 Pendulum-resistance attitude sensor.

The resistance output can be used in a bridge circuit or any of the signal conditioning circuits described for other resistance-type transducers (resistance accelerometer, linear-motion potentiometer, etc.). The output resistance is proportional to a change in attitude from true vertical of the object to which the sensor case is mounted. The position of the wiper arm on the resistance element is such that the output is at 50% of the total resistance when the wiper and pendulum are at true vertical. Typically, the resistance outputs are 0% and 100% at plus and minus 45° respectively. If two pendulum-resistance attitude sensors are mounted on an aircraft as shown in Fig. 2-34, they will produce output resistances that represent pitch and roll attitudes of the aircraft. Note that the pendulum can be damped or undamped as needed.

Pendulum-type sensors can also use the reluctive transduction principle described for accelerometers in Sec. 2-8 and 2-9. In a typical design, the pendulum moves an armature relative to two pairs of inductance-bridge coils placed at right angles to each other. The output of each coil pair provides an amplitude and phase change proportional to the magnitude and direction of attitude change in pitch and roll axes.

Another gravity-reference attitude sensor is shown in Fig. 2-35, which is similar to a carpenter's bubble and is generally known as an electrolytic potentiometer. The sensor consists of a curved glass tube partially filled with a conducting liquid or electrolyte. Electrodes 1 and 3 act as the end terminals, electrode 2 acts as the wiper terminal of a potentiometer. As the tube rotates from a true horizontal position, the resistance to the common electrode increases from one upper electrode and decreases from the other.

An electrolytic-potentiometer sensor is generally connected in a bridge circuit and used with alternating current (to prevent polarization of the electrolyte). The bridge output is a measure of the magnitude and direction of attitude changes from true horizontal.

This same principle can be used for a capacitive gravity-reference sensor. In the capacitive sensor, mercury is used in place of an electrolyte, and

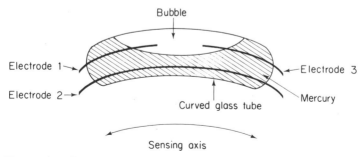

Figure 2-35   Electrolytic-potentiometer gravity-reference attitude sensor.

plates are used instead of electrodes. In effect, the surface of the mercury acts as the common "rotor" plate of a dual-stator capacitor. An a-c bridge circuit is used to convert the capacity changes into voltage changes which, in turn, represent deviations of attitude from true horizontal.

2-10.4 Magnetic-Reference Attitude Sensor

A horizontal reference (to measure yaw) can be established by using the earth's magnetic field in a manner similar to its use in a compass. The reference is a line between an aircraft and magnetic north. In its simplest form, a bar magnet (similar to the magnetic pointer of a compass) is used to move a wiper over a resistance element.

2-10.5 Flow-Stream-Reference Attitude Sensor

Flow-stream sensors are often used on high-speed aircraft and rockets to measure pitch and yaw attitude with reference to local flow stream. Generally, the flow stream is air passing by the vehicle as it moves along. However, the flow stream could be water if the vehicle is used underwater. Sometimes, flow-stream sensors are called *angle-of-attack* transducers, but true angle-of-attack transducers are used to measure attitude in a pitch axis. The angle about a yaw axis is more properly called the *angle of sideslip.*

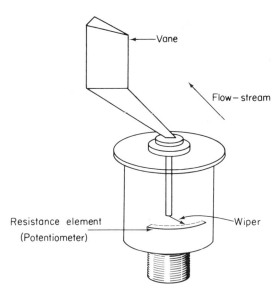

Figure 2-36 Vane-type flow-stream-reference attitude sensor.

There are two basic types of flow-stream sensors: the *vane type* and the *differential pressure* type. The vane-type sensor shown in Fig. 2-36 is essentially a wedge-shaped "weather vane" attached to the wiper of a potentiometer or to the shaft of a synchro transmitter. (Synchros are discussed in Chapter 10.) When installed with the sharp edge of the vane pointing forward, the vane aligns itself with the direction of the airflow passing by the aircraft, and the potentiometer wiper assumes a corresponding position. Depending on the location on the aircraft, the potentiometer (or synchro) output is proportional to either angle-of-attack or angle-of-sideslip.

The differential pressure-type sensor shown in Fig. 2-37 is essentially two tubes mounted on a probe so that the openings of both tubes are flush and parallel. The openings are so located that the difference in pressure between them varies with the direction of air flow relative to an aircraft's axis. For example, if the sensor is used to measure yaw, and there is no yaw (no sideslip), the pressure is equal on both openings. The sensing tubes are connected to a *differential pressure transducer,* such as discussed in Chapter 3, which produces an output resistance or voltage change in proportion to the difference in pressure on the sensor tubes.

2-10.6   Inertial-Reference Attitude Sensors

Operation of inertial-reference attitude sensors is based on the gyro principle. A basic gyro is shown in Fig. 2-38, where a motor-driven wheel or rotor spins on an axle. The rotor is supported by gimbals to allow the spin axis to assume any desired position. Since the gimbal support (ideally) ex-

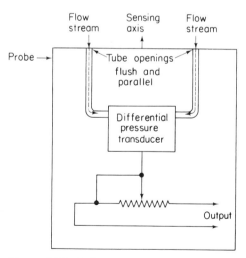

Figure 2-37   Differential pressure-type flow-stream-reference attitude sensor.

erts no torque on the rotor, *the axis of rotation remains fixed in its attitude* as long as the rotor is driven, even if the stand supporting the gimbal rings is moved into a different attitude. If the stand is attached to the vehicle, attitude can be referenced to the rotor axis. This set up follows Newton's first law of motion in reference to inertia. The law states essentially that, unless acted upon by some unbalanced torque, a rotating body will continue turning about a fixed axis.

In the basic gyro of Fig. 2-38, the rotor shaft is supported at each end by a bearing in a frame which, in turn, is free to rotate in bearings attached to a fixed structure whose attitude is to be measured with respect to the rotor spin axis. The frame (or gimbal) rotates about an axis (the gimbal axis) which is at right angles to the rotor spin axis. The fixed structure is usually the gyro case which is attached to a fixed portion of the vehicle and aligned with one of the vehicle axes. Since the position of the spin axis is fixed when the rotor turns, an attitude change of the gyro case (attached to the vehicle) results in an angular displacement between the case and the gimbal shaft. In effect, the spinning rotor axis remains fixed in space, while the gyro and vehicle move. This displacement can be transduced or converted to an electrical signal by any of the means described thus far (variable resistance, inductance, reluctance, etc.). For example, a circular resistance element can be mounted to the case with the wiper arm attached to the gimbal shaft. Note that the gyro illustrated in Fig. 2-38 is a *single-degree-of-freedom* gyro because it is free to rotate in one direction only.

The so-called *free gyro* of Fig. 2-39 is a *two-degree-of-freedom* gyro in which the spin axis may be oriented in any specific attitude. The angular displacement of each of the two gimbals can be converted into an electrical output by suitable resistance elements (or other transduction devices). The two outputs can represent attitude in any two of the three vehicle planes (pitch, yaw, roll), depending on orientation of the gyro within the vehicle. When the gyro is used to establish pitch and roll attitude, with the spin axis

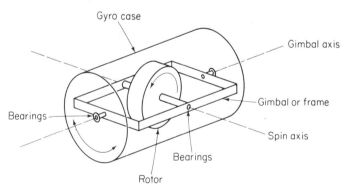

Figure 2-38   Basic (single-degree-of-freedom) gyro.

in a vertical position, it is generally known as a *vertical gyro.* The inner gimbal provides a reference for the pitch attitude, with the outer gimbal providing the roll attitude. When the gyro is used to establish yaw attitude, with the spin axis in a horizontal position, it is generally known as a *directional gyro.* A directional gyro is installed in the vehicle with the outer gimbal parallel to the yaw axis. When the direction of the spin axis is set by a reference relative to the earth's coordinates (magnetic north), such as when the gyro is synchronized or "slaved" to a compass, the directional gyro is known as a *gyro compass.*

One problem with the free gyro of Fig. 2-39 is that the inner gimbal can align itself parallel to the outer gimbal after rotating 90° in either direction. This is known as *gimbal lock,* which causes the inertial attitude reference to be lost. For simple gyros, gimbal lock is eliminated by mechanical stops which prevent the inner gimbal from moving 90°. A *gyro caging* system is often used to prevent gimbal lock and to restore the gyro to normal attitude reference after any occurrence which causes the gyro to lose reference. The typical gyro cage is a solenoid-operated lock which clamps (or cages) the gimbal into some fixed reference position in response to an electrical signal. (Solenoid-operated devices are discussed further in Chapter 9.) Once the gyro is caged and the rotor is up to speed, the gyro can be uncaged by an electrical signal to the solenoid, and the rotor axis will remain fixed at the desired reference attitude. The direction of the spin axis *at the instant of uncaging* determines the reference attitude.

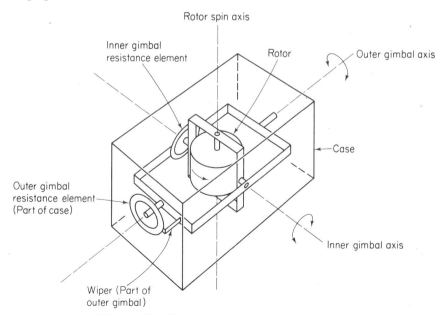

Figure 2-39   Free (two-degree-of-freedom) gyro.

Another problem with gyros is *precession* or *drift* of the spin axis from its intended position. Such drift can be caused by many factors, including unbalance, friction, and magnetic interaction. A well-designed gyro can have a 6° per hour drift of the spin axis. Low-cost designs can have much higher drift rates. There are many techniques used to offset drift. One of the most common is a *torquer system,* which is used to correct constant drift rates or to process the spin axis by given amounts. A torquer is essentially a *rotary solenoid* (Chapter 9), which exerts torque on a gimbal in response to an electrical signal. Torquers are often used in closed-loop systems (Chapter 1) where an error signal, proportional to drift rate, is amplified and applied to the rotary solenoid in a direction to offset the drift.

Another common use of a torquer is in an *erection system* for vertical gyros; such a system is shown in Fig. 2-40. Here, the torquer turns the gimbal clockwise or counterclockwise, as necessary, to keep the spin axis vertical. The torquer solenoid is operated by gravity-sensitive switches similar to those described in Sec. 2-3.3.

Attitude rate can be measured by means of a *rate gyro* such as shown in Fig. 2-41. This gyro is essentially a single-degree-of-freedom gyro but with

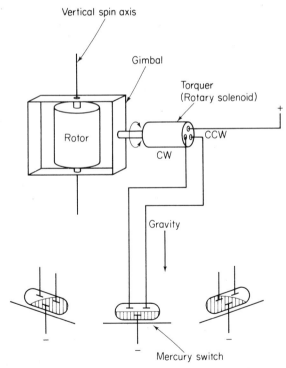

Figure 2-40 Vertical gyro erection system using a rotary solenoid torquer and mercury switches.

restraining springs and dampers. The rate gyro produces an output signal due to gimbal deflections about the output axis (the roll axis in Fig. 2-41), in response to a change in attitude rate about the gyro measuring axis (or input axis). The output signal is taken from a resistance element whose wiper is operated by the gimbal. This output indicates both magnitude and direction of the attitude rate. Generally, the wiper arm is at 50% resistance when the attitude rate is zero. With this arrangement, the wiper moves toward 0% or 100% as the attitude rate varies on either side of zero.

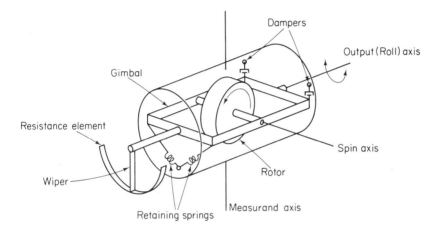

Figure 2-41   Basic attitude rate gyro (for roll axis).

3

FLUID SENSORS

Fluid sensors can be classified by the method of sensing or by the condition they sense. In this chapter, we use the latter classification to cover devices that sense the three most commonly measured conditions of fluid: *flow, pressure,* and *liquid level.*

3-1 FLUID FLOW SENSORS

There are many processes that require a means for measuring fluid flow, and there are two conditions of particular importance. In one case, it is necessary to measure the *total quantity* of the fluid flow; in the other case, the most important factor is the *rate of flow,* or the amount of fluid passing a given point during a given period of time. Before going into the details of specific fluid flow sensors, let us consider the units of measurement and the basic terms common to all flow measurements.

3-1.1 Units of Flow Measurement

Volume flow rate is most commonly expressed in *gallons per minute* (gal/min or gpm) for liquids and in *cubic feet per minute* or *per second* (ft³/min, ft³/s or cfm, cfs) for gases. Volume flow rate can also be expressed in metric units. When volume flow rates are very low, they may be expressed in quantity *per hour.*

Mass flow rate is usually expressed in *pounds (force) per minute* (lb/min) or in equivalent metric units. *Total flow* is expressed in the unit of quantity used for the measurement of the flow rate from which the total flow is derived. *Viscosity* (absolute viscosity) can be expressed in *pounds (force)* per square foot per second (lb-s/ft²).

3-1.2 Basic Flow Measurement Terms

Flow is the motion of a fluid, usually a confined fluid stream. *Flow rate* is the time rate of motion of a fluid expressed as fluid quantity per unit of time. Most flow sensors measure flow rate, even though the term *flow* is used. Also, most flow sensors measure flow of a fluid contained in a pipe or duct.

Total flow is the flow rate integrated over a time interval. *Volumetric flow rate* is expressed as fluid volume per unit of time. *Mass flow rate* is flow rate expressed as fluid mass per unit of time.

Viscosity is the resistance of a fluid to a tendency to flow. *Density* is the ratio of a mass to volume. *Specific gravity* is the ratio of the density of a substance at a given temperature to the density of a substance considered as a standard. Pure distilled water at 4 °C as well as 60 °F is often used as a standard for liquids and solids. Air at 1 atmosphere (14.7 lb/in.² absolute) and at 60 °, 68 °, or 70 °F is used as a standard for gases.

3-1.3 Rate-of-Flow Sensors

One of the most common methods for sensing the rate of fluid flow is to place an obstruction in the path of the fluid, and measure the difference in pressure before and after the obstruction, using a *differential-pressure sensor* (discussed in Sec. 3-2). The obstruction causes a change in fluid pressure which is dependent on the rate of flow. By measuring the difference in pressure, before and after the obstruction, the rate of flow may be determined. The orifice plate and the venturi tube are common devices for this *indirect* rate of flow measurement.

The *orifice plate* shown in Fig. 3-1 consists of a disk with a hole, or orifice, of definite size through its center. The disk is placed within the pipe carrying the fluid. As the fluid encounters the smaller opening, a pressure is built up in front of the plate. As the fluid passes through the orifice and enters the wider pipe, the pressure behind the plate is reduced because a certain amount of pressure is lost overcoming the obstruction. The *difference in pressure* between the front and rear of the plate depends on the size of the orifice and the rate at which the fluid is flowing. The greater the rate of flow, the greater the difference in pressure. As shown in Fig. 3-1, the orifice

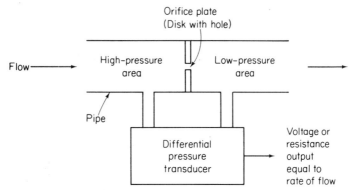

Figure 3-1 Rate-of-flow sensor using orifice plate and differential pressure transducer.

plate is generally used with a differential-pressure transducer such as described in Sec. 3-2.

The *venturi tube* shown in Fig. 3-2 is used more often than an orifice plate. The basic venturi tube consists of a pipe that narrows at a certain point and then widens to its normal diameter. Action of a venturi tube is similar to that of an orifice plate; that is, high pressure is developed where the diameter is greatest, and low pressure where the diameter is smallest. A differential-pressure transducer measure the difference in pressure (which corresponds to flow rate) and converts this difference into an electrical signal equal to flow rate.

Three other common indirect flow rate sensors are shown in Fig. 3-3. The *nozzle* system operates in a manner almost identical to an orifice plate. The *pitot tube* consists of two pressure-measuring tubes, one within the other. The inner tube (point A) receives the full or high pressure of the fluid flow, whereas the other tube (point B) receives the low pressure of the passing fluid. The *centrifugal* system operates by the principle that the outer or top ports will receive the high pressure flow, while the inner or bottom ports receive the low pressure. Again, as for all indirect rate-of-flow systems, the sensor sets up a pressure differential that is proportional to the rate of flow, which is converted into a corresponding electrical signal by a differential-pressure transducer.

There are several methods for obtaining *direct measurement* of fluid flow rate. Two of the most common methods involve a *turbine flowmeter* and a *magnetic flowmeter,* both of which can be considered as rate-of-flow transducers or sensors. That is, with either device, the rate of flow is converted directly into a corresponding electrical signal.

The *turbine flowmeter* shown in Fig. 3-4 consists basically of a rotor mounted axially within a pipe between a set of bearings. As the fluid flows past a set of propeller blades mounted on the rotor, the rotor spins. The

52

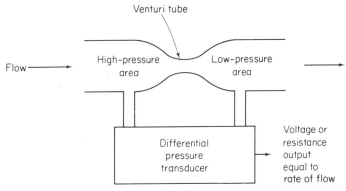

Figure 3-2 Rate-of-flow sensor using venturi tube and differential pressure transducer.

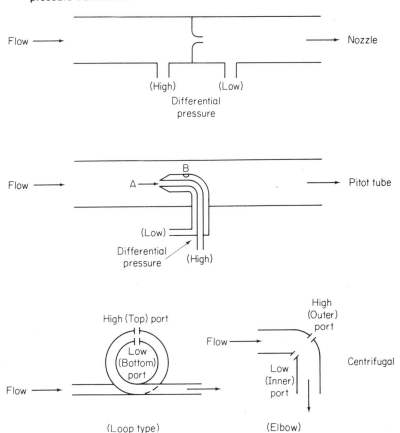

Figure 3-3 Rate-of-flow sensors using nozzle, pitot tube, and centrifugal system.

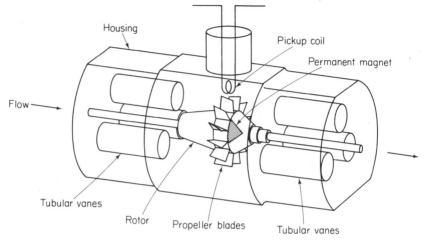

Figure 3-4 Turbine flowmeter.

speed of rotation is directly proportional to the rate of flow through the pipe. The fixed tubular vanes at either side of the rotor reduce the swirling effect of the fluid, an effect that would interfere with the rotation.

A small permanent magnet is mounted within the body of the rotor. As the rotor and magnet spin, a current is induced in a pickup coil within the housing, thus creating a magnetic field around the coil. As each propeller blade of the rotor passes the coil, the magnetic field around the coil is disturbed, thereby producing an electrical pulse in the pickup coil. The frequency of these pulses is directly proportional to the rate of flow through the pipe. By counting the number of pulses during a given period of time, the fluid flow rate can be determined. Counting circuits and devices are described in Chapter 8.

The *magnetic flowmeter* shown in Fig. 3-5 operates by the principle that a voltage is induced in a conductor when the conductor moves through a magnetic field. The magnitude of the voltage depends on the strength of the magnetic field, the length of the conductor, and the speed with which the conductor cuts through the magnetic field. If the strength of the field and the length of the conductor are kept constant, the induced voltage then depends on the speed of the conductor.

In the magnetic flowmeter, the constant magnetic field is provided by a steady direct current flowing through the coil of an electromagnet that surrounds the pipe. The fluid, which must be a conductive material, flows through a nonmetallic section of pipe located in the air gap of the electromagnet. The effective length of the conductor corresponds to the inner diameter of the pipe located in the air gap. Since this diameter is constant, the length of the conductor is constant. Thus, with the length of the conductor and the magnetic field both constant, the induced voltage depends on

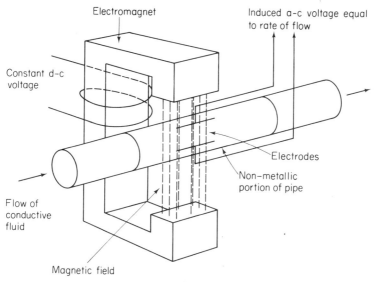

Figure 3-5 Magnetic flowmeter.

the speed with which the conductor cuts through the field (fluid rate of flow). The induced voltage in the conductor is picked up by a means of two electrodes set in opposite sides of the nonmetallic section of pipe. The rate of flow is determined by measuring that voltage.

3-1.4 Quantity-of-Flow Sensors

Most devices that measure quantity of flow operate by accumulating the fluid in a chamber of known capacity and then emptying the chamber. By counting the number of times the chamber is emptied, the total quantity of fluid can be calculated.

The *water meter* shown in Fig. 3-6 is typical of the quantity-of-flow sensors found in most residential water systems. The main element of this water meter is a moving disk which is arranged so that, when tilted to the left, the left-hand part of the chamber is filled with water and access to the exit pipe is shut off. The pressure of incoming water tilts the disk to the right, thus forcing the water to fill the entire chamber and opening access to the exit pipe. When the chamber is again emptied, the disk tilts back to the left and the entire process is repeated.

As the disk tilts from side to side, an arm operates a counting mechanism. If the capacity of the chamber is known, the quantity of water for each count can be determined. The number of counts per given period of time (hour, minute, etc.) then indicates the quantity of flow for the same time interval. The water meter shown in Fig. 3-6 uses a mechanical counter. An electronic counter can also be used in an industrial control system,

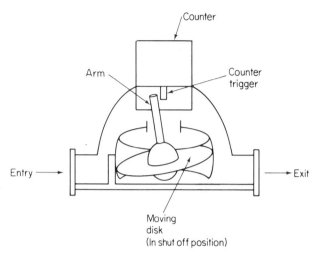

Figure 3-6 Quantity-of-flow sensor (water meter).

whereby the counter receives pulses developed by a switch that is operated each time the disk moves back and forth.

3-2 FLUID PRESSURE SENSORS

There are a number of different devices used for sensing or measuring the pressure exerted by a fluid. Some of the most common include the bellows, Bourdon tube, diaphragm, capsule, and pressure cell or transducer. There are also a number of *differential pressure* sensors. Before going into the details of specific fluid pressure sensors, let us consider the units of measurement and the basic terms common to all pressure measurements.

3-2.1 *Units of Pressure Measurement*

Pressure is usually expressed in units of force per unit area, such as *pounds per square inch* (psi) or kilograms *per square centimeter* (kg/cm²). Linear units of *liquid head* are frequently used to express low pressures, such as *inches of water* at 4 °C (in. H_2O) *centimeters of water* at 4 °C, and *inches of mercury* at 0 °C (in. Hg). Vacuum is measured in *millimeters or microns of mercury* at 0 °C (mm Hg, μ Hg). One mm Hg is the pressure indicated by a column of mercury 1 mm high at 0 °C and at standard gravity. The *torr,* equal to 1 mm Hg at 0 °C, is frequently used for vacuum work. The *atmosphere* is often used for high pressure measurements. One (normal) atmosphere, essentially representing the ambient atmospheric pressure at earth sea level, is the pressure indicated by a 760 mm high column of mercury at 0 °C, at a density of 13.595 g/cm³ and an acceleration due to gravity of 980.665 cm/s².

3-2.2 Basic Pressure Measurement Terms

Pressure is force acting on a surface and is measured as *force per unit area* exerted at a given point. *Absolute pressure* is measured relative to zero pressure and is expressed as *pounds per square inch, absolute* (psia). *Gage pressure* is measured relative to ambient pressure, and is expressed as *pounds per square inch, gage* (psig). At sea level, 0 psig = 14.7 psia (approximately). *Differential pressure* is the difference in pressure between two points of measurement and is measured relative to a reference pressure or a range of reference pressures. Typically, differential pressure is expressed as *pounds per square inch, differential* (psid).

Standard pressure is a pressure of 1 normal atmosphere (Sec. 3-2.1). *Static pressure* is the pressure of a fluid, exerted normal to the surface along which the fluid flows. A fluid can be liquid or gaseous. The static pressure of a moving fluid is measured normal to the direction of flow. *Impact pressure* is the pressure in a moving fluid exerted parallel to the direction of flow, due to flow velocity. *Stagnation pressure* (also called *ram pressure* or *total pressure*) is the sum of static pressure and impact pressure.

Head is the height of a liquid column at the base of which a given pressure would be developed. *Partial pressure* is the pressure exerted by one constituent of a mixture of gases. *Vacuum* is pressure reduced to a pratically attainable minimum in a volume or region of space. Perfect vacuum is zero absolute pressure and the complete absence of any matter.

3-2.3 Bellows Pressure Sensors

The basic bellows pressure sensor shown in Fig. 3-7(a) consists of a cylindrical metal box with corrugated walls of thin, springy material such as brass, phosphor-bronze, or stainless steel and is generally used where the pressures involved are low.

Pressure inside the bellows tends to extend its length. This tendency is opposed by the springiness of the metal, which tends to restore the bellows to its normal size. Pressure on the outside of the bellows tends to reduce its length. This tendency is also opposed by the springiness of the metal.

For low pressures, the springiness of the metal will restore the bellows. For larger pressures, a spring inside or outside the bellows may be used to reinforce the springiness in its tendency to restore the bellows to normal.

As shown in Fig. 3-7(a), pressure inside the bellows extends its length. Since the bottom end of the bellows is fixed, the travel of the rod connected to the top end of the bellows is directly proportional to the pressure within. The motion of this rod can be used to move the slider of a resistance element or other linear-motion transducer (Chapter 2). The electrical output signal from the resistance element is then proportional to the fluid pressure applied to the bellows.

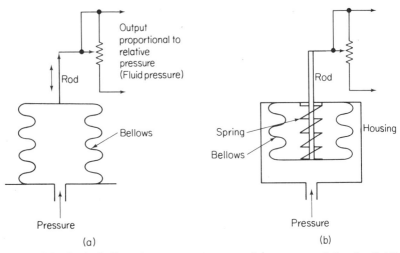

Figure 3-7 Basic bellows-type pressure sensor to measure relative (or fluid) pressure.

The pressure sensor shown in Fig. 3-7(b) has the basic bellows enclosed in a housing. Pressure in the housing tends to compress the bellows, shortening its length. This action is opposed by the springiness of the metal, reinforced by an internal spring. The top of the bellows is fastened to the top of the housing. As pressure is increased, the free bottom of the bellows moves toward the top. A rod, fastened to the bottom and passing through an opening in the top of the bellows and housing, transmits this motion to the external device (such as a linear motion transducer).

Note that fluid pressure within the basic bellows of Fig. 3-7(a) tends to expand the bellows, whereas atmospheric pressure (outside) tends to contract the bellows. The reverse is true of the bellows shown in Fig. 3-7(b), where fluid pressure produces contraction and an increase in atmospheric pressure produces expansion of the bellows. With either system, the bellows senses the *relative* pressure, or difference between the absolute fluid pressure and the atmospheric pressure. (Generally, the term *fluid pressure* is taken to mean the relative pressure.)

There are circumstances where it is necessary to measure the *absolute* pressure of a fluid. A pressure sensor for such use is shown in Fig. 3-8. Here, a *sensing* bellows is enclosed in a housing made airtight by means of a *sealing* bellows. All the air in the housing is removed, leaving a vacuum. Thus, there is no atmospheric pressure on the sensing bellows to counter the action of the fluid pressure. The sealing bellows also permits free movement of the arm actuated by the sensing bellows.

The bellows element and its transducer can also be combined into a single unit as shown in Fig. 3-9. In Fig. 3-9(a), the movements of the

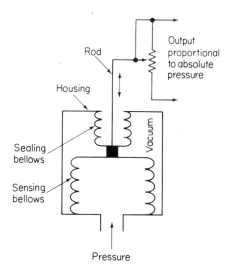

Figure 3-8 Bellows-type pressure sensor to measure absolute pressure.

bellows, resulting from variations in the fluid pressure applied, cause a slider to move over a resistance element located within the housing. The resistance output signal is thus proportional to the fluid pressure. In Fig. 3-9(b), the movements of the bellows cause a magnetic core to be moved in or out of a coil mounted within the housing. The inductance of the coil, which is the output signal, is proportional to the fluid pressure causing the movement.

3-2.4   Bourdon-Tube Pressure Sensors

The basic Bourdon tube shown in Fig. 3-10 consists of a tube with a flattened cross section made of a springy metal, such as phosphor-bronze or stainless steel, in the form of an incomplete circle. As the fluid pressure inside the tube in increased, the tube tends to straighten. As the pressure is reduced, the tube tends to return to its curved form. The position of the free end of the tube thus varies with changes in the fluid pressure.

The motion of the free end can be used to operate a motion-sensing transducer to produce an electrical output. For example, the free end of the Bourdon tube can be coupled to a linear variable differential transformer to produce a voltage signal that is proportional to the fluid pressure or to an angular motion potentiometer to produce a resistance signal similarly proportional to the pressure.

As shown in Fig. 3-10, there are several variations of the Bourdon tube. In one variation, the tube is wound in a flat spiral and is used for medium-

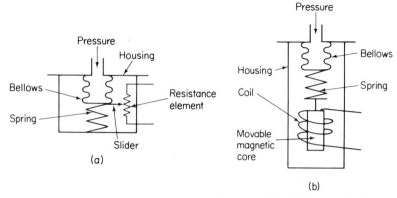

Figure 3-9 Bellows element and transducer combined into a single-unit pressure sensor.

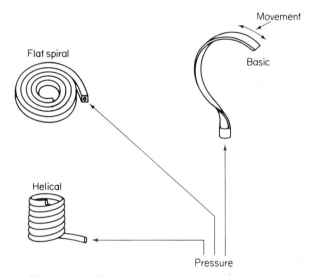

Figure 3-10 Basic Bourdon tube pressure sensors.

range pressures. In another variation, the tube is wound in a long helical spiral and is used for high pressures. The basic tube of Fig. 3-10 is generally used for low pressures. However, with proper materials and heavy wall thickness the basic tube can be used for medium pressures. The advantage of the flat spiral and helical tubes over the basic tube is that the travel of the free end is greater for a given pressure.

3-2.5 Diaphragm and Capsule Pressure Sensors

The basic diaphragm sensor shown in Fig. 3-11(a) is a circular disk of thin springy metal firmly supported at the rim. As pressure is applied to one side of the diaphragm, the center is bent away from the pressure. This bend-

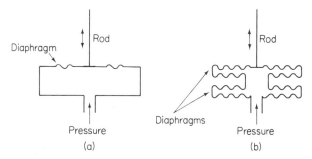

Figure 3-11 Diaphragm and capsule pressure sensors: (a) basic diaphragm; (b) capsule.

ing motion is opposed by the springiness of the metal, and the pressure of the center is directly proportional to the pressure in the case.

The diaphragm surface is usually corrugated to increase the effective area, or it can be flat. Whichever construction is used, motion of the diaphragm can be linked to a motion-sensing transducer (resistance element, etc.) to produce an output signal that is proportional to pressure applied on the diaphragm.

Often, two or more diaphragms are joined to form a *capsule* as shown in Fig. 3-11(b). Increasing the pressure in the capsule causes it to expand; decreasing the pressure causes the capsule to contract. Since all of the diaphragms of the capsule act together, the travel of the arm for a given pressure is greater than that of a single diaphragm.

The diaphragm sensor may be combined with a transducer in a single unit as shown in Fig. 3-12. A fixed metal plate is mounted inside a rigid housing. The diaphragm is mounted in front of the plate in such a way as to divide the housing into two compartments. The diaphragm and fixed plate form the plates of a capacitor. When pressure is applied, the diaphragm is bent closer to the fixed plate. As a result, capacitance is increased. When pressure is relieved or decreased, the springiness of the metal forces the diaphragm back to its original position, thus reducing capacitance. The variable capacitance (which varies with pressure changes) is the output signal. Generally, capacitance output forms one leg of an a-c bridge circuit as shown in Fig. 3-12.

3-2.6 *Pressure Cell Sensors and Pressure Transducers*

The *pressure cell* shown in Fig. 3-13 consists of a thin-walled, flattened metal tube made of springy material. One end of the tube is sealed, and pressure is applied through the open end. The pressure in the tube tends to inflate it, a tendency that is opposed by the springiness of the metal. As a result, a strain is set up in the wall of the tube. This strain is directly proportional to the pressure.

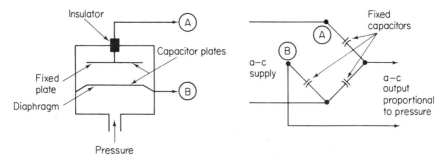

Figure 3-12 Diaphragm-type pressure sensor using capacitor transducer output.

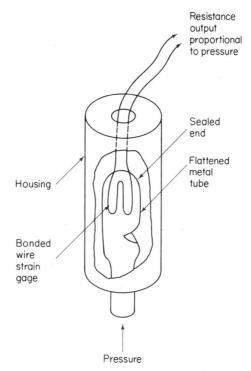

Figure 3-13 Basic pressure cell.

A bonded wire strain gage, such as discussed in Sec. 2-5, is cemented to the wall of the tube. As the tube expands, the strain in the wall stretches the wire of the gage, increasing the resistance. Decreases in pressure decrease resistance. The resistance output, which corresponds to fluid pressure, can be used in the usual manner (as one leg of a bridge circuit).

The *pressure transducer* shown in Fig. 3-14 is similar to a pressure cell. Pressure transducers are often used to measure gas pressures such as those within the cylinder of an internal-combustion engine. The diaphragm end of

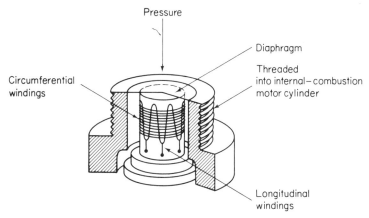

Pressure

Diaphragm

Threaded
into internal–combustion
motor cylinder

Circumferential
windings

Longitudinal
windings

Figure 3-14   Pressure transducer for measurement of gas pressures.

the transducer is inserted through a threaded hole in the side of the cylinder. Pressure transducers are also used in automobile radiators.

The pressure to be measured is imposed on the diaphragm and as the pressure increases, the cylindrical strain tube decreases in length while increasing in diameter. These dimensional changes are detected by the wire strain gages bonded to the strain tube. The resistance of the circumferential winding increases and the longitudinal winding decreases with an increase in pressure. These resistance changes, which are proportional to fluid pressure changes, are used as the output signal or signals in the normal manner.

3-2.7   Differential-Pressure Sensors

As its name implies, a differential-pressure sensor senses the *difference in pressure* between two fluids or between different portions of the same fluid (such as described in Sec. 3-1.3). The three most common differential pressure transducers are the *manometer, differential-bellows,* and *differential-capsule* sensors.

The basic *manometer* shown in Fig. 3-15 consists of a U-shaped tube containing some liquid. Laboratory and test manometer tubes are made of clear glass or plastic, with an inch or centimeter scale etched on the surface so that the height of the liquid can be measured directly. The liquid used in most lab manometers is mercury.

With no pressure applied, the mercury column in each arm of the tube is at the same height. If the pressure applied to arm *B* is greater than the pressure to arm *A,* the mercury is forced down in arm *B* and up in arm *A.* The distance or height between the tops of the columns (designated as *H* in Fig. 3-15) is proportional to the *difference* between the two pressures. If the pressure applied to arm *B* is less than the pressure at arm *A,* the mercury is forced down in arm *A* and up in arm *B.* Again, the distance between the

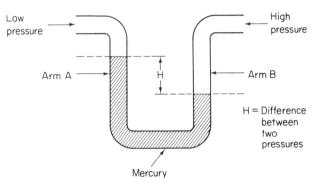

Figure 3-15 Basic manometer.

tops of the columns is proportional to the pressure differential. In this way, the sensor indicates which of the pressures is greater (by movement of the mercury) and the differential in pressures (as indicated by H).

The basic manometer principle can be used to provide an electrical signal. There are several methods for such conversion. In one method the linear variable differential transformer (discussed in Sec. 2-2) is used as a secondary transducer to translate the motion of the mercury in the tube to an electrical signal as shown in Fig. 3-16. Here, a float riding on top of the mercury column carries a rod attached to the core of the transformer. The windings are on the outside of the manometer tube. Since the tube is made of glass or plastic, it does not affect action of the transformer. Changes in the height of the mercury column produce variations in the output from the transformer. The change in amplitude and phase of the output correspond to the rise and fall of the column and thus indicate the changes in pressures (or differential pressure) applied to the manometer.

Two bellows elements may be combined to form a differential-pressure sensor as shown in Fig. 3-17. The bellows elements are linked by a rod between them. Pressure *A*, applied to the left-hand bellows, moves the rod to the right. Pressure *B*, applied to the right-hand bellows, moves the rod to the left. The resultant movement of the rod is proportional to the differential between pressure *A* and pressure *B*. As the rod moves, an arm transmits the motion to an external transducer (such as a resistance element).

The *capsule* may also be used as a differential-pressure sensor as shown in Fig. 3-18. Pressure *A* is applied on the outside of the actuating capsule, tending to collapse the capsule. Pressure *B* is applied to the inside of the capsule, and tends to expand the actuating capsule. Since the bottom of the capsule is fixed, the motion of the top is the result of the differential pressure. This motion is transmitted to a pivoted arm whose movement can be used to actuate the slider of a resistance element or other transducer. Note that the sealing bellows is used to keep the housing airtight and yet permit movement of the arm.

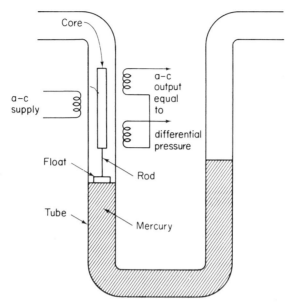

Figure 3-16 Basic manometer combined with LVDT to provide an electrical output equal to differential pressure.

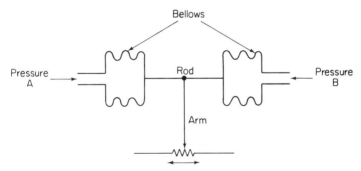

Figure 3-17 Differential-pressure transducer with two opposing bellows elements.

3-3 LIQUID LEVEL SENSORS

There are a number of different devices used for sensing or measuring the level of a liquid in a tank or other container. Some of the most common include the pressure sensor, float-operated resistance element, and capacitive or conductive level sensors.

Any of the fluid-pressure sensors described in Sec. 3-2 can be used to measure liquid level by placing the sensor at the bottom of a tank or container as shown in Fig. 3-19. The pressure the liquid exerts on the bottom of the tank depends, in part, on the height of the liquid. Thus, by measuring the pressure on the bottom, the level within the tank can be determined.

The *float-operated resistance element* shown in Fig. 3-20 is the most common liquid level sensor. As the liquid rises in the tank, the float (typically a hollow metal or plastic ball) is raised, and its arm causes the slider to move over the resistance element so that the resistance is increased. As the liquid level falls so does the float, and the resistance is decreased. Thus, the resistance is proportional to the level of the liquid in the tank.

The *capacitive liquid-level sensor* shown in Fig. 3-21 is the next most common liquid level sensor. This device consists of an insulated metal electrode firmly fixed near and parallel to the metal wall of the tank. If the liquid is nonconductive, the electrode and tank wall form the plates of a capacitor, and the liquid between them acts as a dielectric.

The capacitance depends, in part, on the height of the dielectric (or liquid) between the plates. The greater the height, the larger the capacitance and vice versa. Thus, the capacitance is directly proportional to the level of the liquid in the tank.

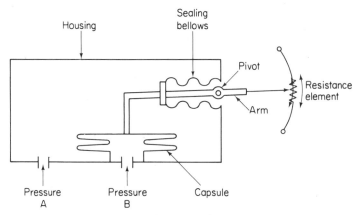

Figure 3-18 Differential-pressure transducer with capsule and sealing bellows.

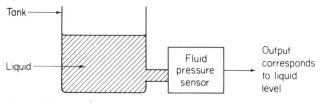

Figure 3-19 Liquid-level sensor using fluid pressure measurement.

The *conductive liquid-level sensor,* also shown in Fig. 3-21, is similar to the capacitive sensor, except that the conductive sensor is used where the liquid is conductive. The conductive liquid acts as a variable resistance between the rod and tank wall. The resistance depends on the height of the liquid between the rod and tank.

Where the tank is not made of metal, two or four parallel insulated rods, held a fixed distance apart, are used as the capacitor plates (or resistance element electrodes).

In addition to these common types of level sensors, there are some sen-

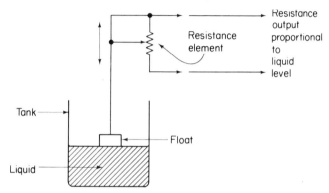

Figure 3-20 Float-operated liquid-level sensor.

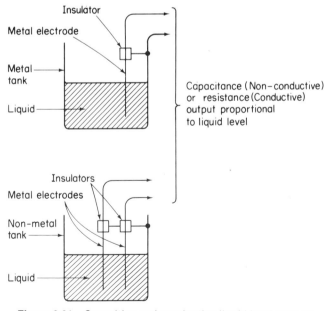

Figure 3-21 Capacitive and conductive liquid-level sensors.

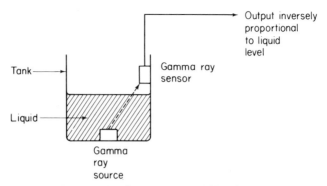

Figure 3-22 Gamma-ray liquid-level sensor.

sors that use unique methods for determining the level of liquid in tanks. The *gamma-ray* sensor shown in Fig. 3-22 is an example of a sensor that is used where it is not practical to apply an electrical current to the liquid (as is required for the capacitive and conductive sensors). (Gamma rays, which are electromagnetic waves similar to X rays, are discussed in Chapter 5.)

In the sensor of Fig. 3-22, a source of gamma rays is placed at the bottom of the tank. A gamma-ray sensor, such as the Geiger-Muller tube discussed in Chapter 5, is mounted outside the tank at a point near the top. Since the gamma rays can penetrate the walls of the tank, the sensor detects the rays coming from the bottom of the tank. The greater the intensity of the rays, the greater the output of the sensor. The output is fed to a counting device. Maximum reception of the gamma rays occurs when there is only air between the source and the sensor. As the level rises, a certain amount of liquid comes between the two and, since the liquid absorbs some of the rays, fewer rays reach the sensor, causing the sensor's output to decrease. The higher the level of absorption, the lower the sensor output. Thus, the sensor output is inversely proportional to the liquid level.

3-3.1 Liquid-Level Limit Sensors

For some applications, it is necessary to sense when a liquid flowing into a tank reaches a predetermined level. When that level is reached, a control circuit may shut off or otherwise regulate the liquid flow. The simplest of such devices is the *float switch* shown in Fig. 3-23. In a typical system, the switch is mounted near the top of the tank. A float connects to the arm of the switch by means of a rod. When the liquid level reaches the predetermined level, the float rises enough so that the rod can push the switch arm as far up as it can go. With the arm in that position, the switch is closed and so is the control circuit. When the liquid level drops, so does the float. The arm is pulled down, and the switch is opened as is the control circuit.

When the liquid is nonflammable, a simple level-limit sensor can be made by using an electrode as shown in Fig. 3-24. The electrode is set in the side of the metal tank at the desired level. When the liquid reaches that level, contact is made between the liquid and the electrode to complete the control circuit. When the level drops below the electrode, contact is broken and the control circuit is opened. For this circuit it is assumed that the tank is made of metal and makes contact with the liquid. Where there is no such contact (glass or plastic tanks) the circuit is completed by extending another electrode into the liquid through the tank wall near the bottom.

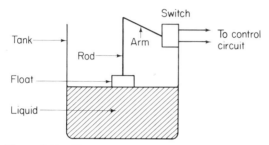

Figure 3-23 Float-operated liquid-level limit sensor (and control switch).

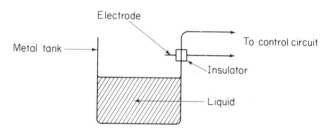

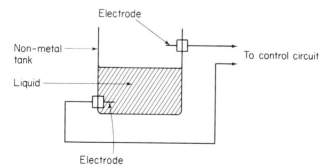

Figure 3-24 Simple liquid-level-limit sensors for nonflammable liquid.

4

MOISTURE AND HUMIDITY SENSORS

Moisture and humidity sensors can be classified by the method of sensing or by the conditions sensed. The two most common instruments used for moisture and humidity measurement are the *hygrometer* and the *psychrometer,* each using a different method of sensing. *Dew point* is one of the most important conditions to be sensed. Before going into the details of specific sensors, let us consider the basics of moisture and humidity measurement.

4-1 MOISTURE AND HUMIDITY MEASUREMENT BASICS

Air is a mixture of several gases including water vapor. The amount of water vapor in a given quantity of air is called the *humidity*. Generally, humidity is measured in grains of water vapor per cubic foot of air. (There are 7000 grains per pound or 437.5 grains per ounce.) The total amount of water vapor a given quantity of air can absorb varies with air temperature. The higher the temperature, the more water vapor the air can hold. For example, at 0 °F a cubic foot of air can absorb a maximum of about 0.5 grain. At 100 °F, the same cubic foot can absorb about 20 grains.

This does not mean that air at a given temperature always contains a certain amount of water. Air may contain less than the maximum amount absorbable. Thus, a cubic foot of air at 100°F may contain only 10 grains of water vapor. Under such conditions, the air is said to contain 50% of the water that could be absorbed. Another way of saying this is that the *relative humidity* of the air is 50%. In effect, the relative humidity is the percentage of water vapor actually present in the air compared with the maximum amount that could be present at a given temperature.

If the air contains less water vapor than the maximum amount it can hold at a given temperature (if the relative humidity is less than 100%), it is possible for the air to absorb more water. The lower the relative humidity of the air, the faster it can absorb water vapor. If the air contains the maximum amount of water vapor for a given temperature (relative humidity is 100%) the air can absorb no more water. Any attempt to force more water vapor into the air causes the excess to condense into water.

Air conditioners control the moisture content of the air in a room by using this principle. If the moisture content is too low, the air is heated, thus reducing the relative humidity. The air is then able to absorb water vapor faster. If the moisture content is too high, the air is cooled below the point where the relative humidity is 100%. The excess moisture is condensed into water.

Control of the moisture content of the air is quite important for many industrial processes. For drying operations, the lower the relative humidity of the air, the faster drying takes place. For some processes, quick drying is not desirable so the relative humidity is kept high. For other processes, it is desirable to maintain a constant relative humidity.

4-1.1   Units of Moisture
and Humidity Measurement

Moisture is expressed in *percent by weight* (either with respect to total weight or to dry weight) or in *percent by volume*. *Relative humidity* is expressed in *percent*. *Absolute humidity* is typically measured in *grams per cubic meter*. *Humidity mixing ratio* is usually expressed in *pounds per pound, grams per pound,* or *grams per kilogram*. *Dew point* is shown in degrees (°F or °C).

4-1.2   Basic Moisture
and Humidity Measurement Terms

Moisture refers to the amount of liquid absorbed by a solid. Moisture can also mean the amount of water absorbed in a liquid. *Humidity* is a measure of the water vapor present in a gas. Humidity is usually measured as absolute humidity, relative humidity, or dew-point temperature. *Absolute humidity* is the mass of water vapor present in a unit volume. *Specific humidity* is the ratio of the mass of water vapor contained in a sample (of moist gas) to the mass of the entire sample.

Relative humidity is the ratio (expressed in percent) of the water-vapor pressure actually present to water-vapor pressure required for saturation at a given temperature. Relative humidity is always temperature-dependent.

The *dew point* is the temperature at which the saturation water-vapor pressure is equal to the partial pressure of the water vapor in the atmosphere. Any cooling of the atmosphere below the dew point will produce water condensation. The relative humidity at the dew point is 100%. The dew point can also be defined as the temperature at which the actual quantity of water vapor in the atmosphere is sufficient to saturate that atmosphere with water vapor (the temperature at which dew will form).

4-2   HYGROMETER-TYPE SENSORS

A hygrometer is an instrument which measures humidity *directly*. Generally, a hygrometer is calibrated in terms of relative humidity but can also be used to indicate absolute humidity. There are five basic types of hygrometers: *resistive, dielectric-film, mechanical-displacement, oscillating-crystal,* and *aluminum-oxide.* At one time, the mechanical-displacement type was the most popular, but it has been replaced by the resistive type.

4-2.1   *Resistive Hygrometer Sensors*

The most basic method of determining the moisture content of solids in granular or powdered form is to place two contacts, a fixed distance apart, into the material. The resistance of most materials varies inversely with the amount of moisture they contain. Thus, by knowing the resistance of a dry material and measuring the resistance of a moist material, an indication of the moisture content can be obtained. An example of this basic system is used in textile and paper processing control systems. Many materials have a high value of resistance when they are dry but a much lower value when they are wet or moist. The moisture content of such materials may be continuously monitored as they pass between a pair of metal rollers that make contact with the materials. One of the rollers is electrically grounded, and the other is connected to a source of direct voltage. As material passes between the rollers, the resistance to the flow of current between the rollers varies inversely with the voltage. The amount of current is then an indication of the moisture content. (Voltage, current, and resistance measurements are discussed further in Chapter 8).

The most popular resistive hygrometer sensors are those in which a variation of *ambient relative humidity* produces a variation in their resistance. This resistance change occurs in certain materials such as hygroscopic salts and carbon powder. These materials are usually applied as a film over an insulating substrate and are terminated by metal contacts as

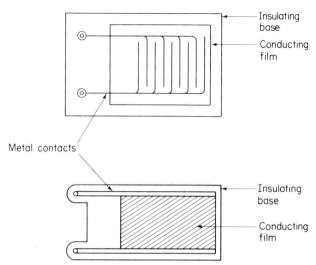

Figure 4-1 Resistive hygrometer sensors (chemical humidity —sensing elements).

shown in Fig. 4-1. However, resistive hygrometer elements are also found in cylindrical form. *Lithium chloride* is the most common hygroscopic salt. The resistance of lithium chloride changes when it is exposed to variations in humidity. The higher the relative humidity, the more moisture the lithium chloride absorbs and the lower its resistance. Thus, the resistance of the sensor may be used as a measure of the relative humidity of the air to which it is exposed. Lithium chloride sensors are used with alternating current, since direct current tends to break down the chemical to its lithium and chlorine atoms.

4-2.2 Dielectric-Film Hygrometer Sensors

The hygroscope-film technique described for the resistive sensors has also been used in certain capacitive humidity-sensing elements, where the film acts as part of the dielectric between two capacitor plates. Changes in humidity produce changes in the dielectric and corresponding changes in capacitance. Dielectric-film hygrometers are not in general use except for some very specialized laboratory applications.

4-2.3 Mechanical-Displacement Hygrometer Sensors

The mechanical-displacement technique was used in one of the original hygrometers. Such sensors respond to certain moisture-absorbing substances that undergo dimensional changes dependent on the amount of moisture absorbed. Human hair is such a substance and is used in a device known as the *hair hygrometer* shown in Fig. 4-2. A number of strands of hair are fastened at one end and connected at the other end through a mechanical linkage to some device, such as the pen of a recorder or the pointer of a meter. Because of the hygroscopic property of hair, an increase in relative humidity causes the hair to increase in length. The spring, acting on the arm, forces the point to move toward the high end of the scale. A decrease in relative humidity causes the hair to contract, thus forcing the pointer toward the low end of the scale. This scale may be calibrated in units of relative humidity (percentage). Mechanical-displacement hygrometer sensors have, in general, been replaced by the resistive sensor (Sec. 4-2.1).

4-2.4   Oscillating-Crystal Hygrometer Sensors

The quartz crystal, discussed in Chapters 2 and 3, can be used as a moisture sensor when its surface is coated with a hygroscopic material (typically a polymer) as shown in Fig. 4-3. The crystal is used in an oscillator circuit. The mass of the crystal varies with the amount of water absorption on the coating, which changes the oscillator frequency. Thus, the frequency becomes a measure of humidity surrounding the crystal.

4-2.5   Aluminum-Oxide Hygrometer Sensors

The electrical properties of anodized aluminum can be used as a moisture-sensing element. A typical sensor of this type consists of an aluminum rod, needle, or strip which has been anodized (Fig. 4-4). When aluminum is anodized, a thin layer of aluminum oxide is formed on the surface, so that the surface consists of many pores. A thin metal coating (generally gold but possibly aluminum) is then deposited over the anodized surface. The outer electrode or contact is formed by the metal coating. The inner electrode is formed by the aluminum base.

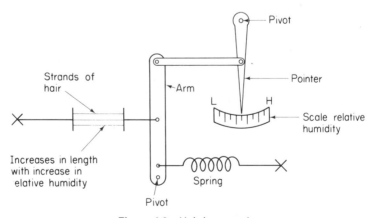

Figure 4-2   Hair hygrometer.

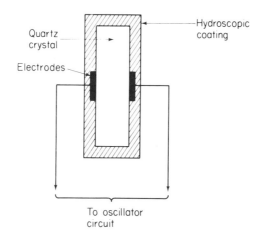

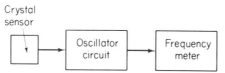

Figure 4-3 Oscillating-crystal hygrometer sensor.

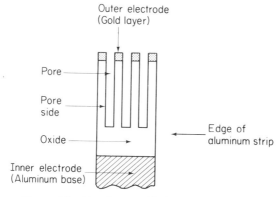

Figure 4-4 Aluminum-oxide hygrometer sensor.

Water vapor passes through the thin metal coating into the pores and fills the pores by an amount corresponding to the vapor pressure of water in the atmosphere surrounding the element. The amount of water absorbed by the pores changes the impedance (both capacitance and resistance) as measured between the electrodes. (Impedance is discussed further in Chapter 8.) The output of the sensor is thus an impedance that varies in relation to humidity. Generally, the impedance is measured by an a-c bridge circuit, as discussed in Chapter 8.

4-3 PSYCHROMETER-TYPE SENSORS

The psychrometer is a humidity-measuring instrument which uses one *wet-bulb thermometer* and one *dry-bulb thermometer*. For control system applications, the thermometers are generally resistive or thermocouple, as discussed in Chapter 6. The dry-bulb thermometer measures ambient temperature. The wet-bult thermometer measures temperature reduction due to evaporative cooling. A wick, a porous ceramic sleeve, or a similar device saturated with water is placed in physical contact with the "bulb" (sensing portion of the thermometer) of the wet-bulb thermometer to keep the bulb moist.

Relative humidity is determined from the two temperature readings and a reading of the barometric pressure, usually by means of a *psychrometric table*. At any given ambient temperature (dry-bulb reading), the relative humidity decreases as the difference between dry-bulb and wet-bulb readings increases. This temperature difference is referred to as the *wet-bulb depression*.

4-3.1 Typical Psychrometer System

Figure 4-5 shows the elements of a typical psychrometer system. These elements include two resistive-type thermometers. As discussed in Chapter 6, such thermometers produce a resistance output which varies with

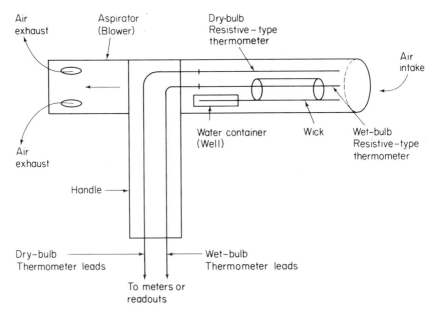

Figure 4-5 Portable psychrometer system.

temperature and thus provides an indication of temperature. The wet-bulb thermometer sensing element is covered by a cloth wick that dips into a well of water. The wick is kept moist by capillary action, which draws water from the well. The dry-bulb thermometer sensing element is near but not in contact with the wick.

The system shown in Fig. 4-5 operates on the principle of cooling by evaporation. When a thin film of water is brought into close contact with air, a portion of it evaporates, and the temperature of the remaining water is lowered due to the heat required by the evaporation process. This evaporation and lowering of the temperature is dependent on the ability of the air to absorb moisture. Thus, the lower the relative humidity, the faster the evaporation and the lower the temperature of the remaining water.

Moving air must be used; if not, the moisture-laden air will blanket the thermometers, thus producing erroneous readings. (The dry-bulb sensor can become wet if the air is not moving.) In the system of Fig. 4-5, the moving air is provided by a fan or blower (also known as an *aspirator* in this application). The dry-bulb thermometer is not affected by the moving air and registers the ambient temperature in the normal manner. However, the wet-bulb thermometer registers a lower temperature due to the evaporation of the water in the wick around the sensor. Since this cooling effect is a function of the relative humidity of the air, the relative humidity may be determined by comparing the readings of the two thermometers. Figure 4-6 is a typical psychrometric table that shows the relative humidity at selected temperatures. Note that relative humidity is dependent on both *ambient*

Difference between dry-bulb and wet-bulb readings, °F				
Dry-bulb reading °F	1	10	20	30
40	92	—	—	—
50	93	38	—	—
60	94	48	6	—
70	95	55	20	—
80	96	61	29	4
90	96	65	36	13
100	96	68	42	21
110	97	70	46	27

Figure 4-6 Partial psychrometric chart showing the relationships between relative humidity, ambient temperature (dry-bulb reading) and difference between dry- and wet-bulb readings.

temperature (dry-bulb reading) and the *difference* between dry- and wet-bulb readings.

4-4   DEW-POINT SENSORS

Dew-point measurement is a measure of the temperature of a surface at the instant when moisture (dew) is first precipitated on the surface as the surface is cooled. For example, assume that the temperature is decreasing and dew forms just as the temperature reaches 25 °C. The dew point is, then, 25 °C. Thus, the dew point is a discrete temperature. In control applications, the dew point is determined by artifically lowering the temperature of the surface and noting the temperature at which dew (or frost) first condenses on the surface. A dew-point sensor must, therefore, perform the function of temperature sensing as well as noting the change from vapor to liquid (instant-of-condensation-sensing function).

4-4.1   Temperature-Sensing Function

The temperature at which condensation first occurs upon cooling of a surface is sensed by resistive or thermoelectric sensing elements (as discussed in Chapter 6). For practical applications, the temperature sensor is kept as close to the surface as possible. Ideally, the temperature sensor is mounted flush with the surface and kept very small in size, since its intended use is to measure the temperature of the condensate itself.

4-4.2   Instant-of-Condensation-Sensing Function

This function is usually provided by a thin disk or plate with a smooth surface which is closely coupled thermally to a cooling element and a condensation detector. Figure 4-7 shows the basic elements of a *resistive-type* dew-point sensor. The condensation surface is an insulating glass and epoxy disk. The condensation sensor is a resistive element in the form of a metal conducting grid imbedded in the disk. The temperature-sensing function is provided by a thermocouple (Chapter 6) also imbedded in the disk.

During use, the disk is cooled from the side opposite the sensors as shown in Fig. 4-7. When the dew point is reached, the resistance of the conducting grid changes drastically (usually unbalancing a bridge circuit). The temperature-sensing thermocouple provides an indication of the temperature at the dew point, as well as before and after the dew point is reached when the disk is cooled.

Figure 4-8 shows the basic elements of a *photoelectric-type* dew-point sensor. Here, the condensation surface is a mirror or mirror-like metal reflector, and the temperature-sensing function is provided by a thermo-

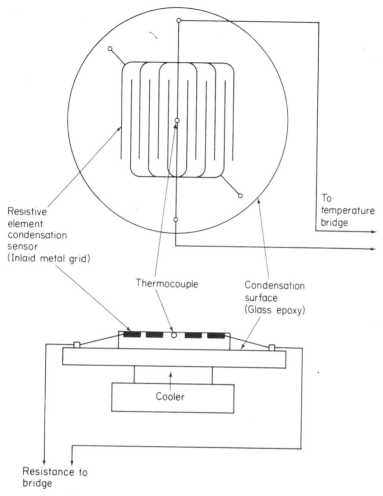

Resistive
element
condensation
sensor
(Inlaid metal grid)

Thermocouple

Condensation
surface
(Glass epoxy)

To
temperature
bridge

Cooler

Resistance to
bridge

Figure 4-7 Resistive-type dew-point sensor.

couple. A constant light source is trained on the mirror surface, and the reflected light is sensed by some form of light sensor (as discussed in Chapter 5).

Again, the disk is cooled from the side opposite the sensors. When the dew point is reached the mirror surface is fogged, and the light sensor output is drastically changed (unbalancing a bridge or similar circuit). Again, the temperature-sensing thermocouple provides an indication of the temperature at the dew point.

Figure 4-9 shows the basic elements of a *radiation-* or *nuclear*-type dew-point sensor. Here, an alpha- or beta-particle radiation source (Chapter 5) is

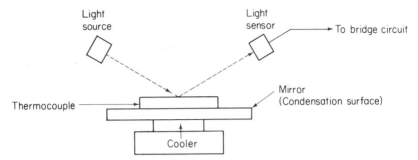

Figure 4-8 Photoelectric-type dew-point sensor.

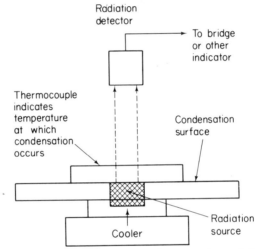

Figure 4-9 Radiation or nuclear type dew-point sensor.

located flush with the condensation surface, and a radiation detector (Chapter 5) senses the drop in particle flux when condensation forms over the radiation source (dew point). Again, a temperature sensor indicates the temperature at the dew point.

5

LIGHT AND RADIOACTIVITY SENSORS

Both light and radioactivity are a form of radiation. Thus, the devices which sense light and radioactivity are, in effect, radiation sensors. In this chapter we consider sensors that can sense and measure the intensities of electromagnetic radiations such as light and X rays and nuclear radiations, so-called because they come from the nuclei of radioactive atoms. These nuclear radiations consist, for the most part, of beams of alpha, beta, and neutron particles and gamma rays. We start with a discussion of light sensors.

5-1 LIGHT SENSORS

Light represents a small portion of a whole spectrum of electromagnetic waves, which differ from one another in wavelength. This spectrum includes, in order of decreasing wavelength, radio waves, light waves, X rays, and gamma rays. The category "light" includes *infrared, visible light,* and *ultraviolet* rays. Infrared light has the longest wavelengths and ultraviolet light the shortest.

A *light sensor* has many applications as a detector of the presence or absence of light and for measuring the intensity of light. Basically, a light sensor is a *photocell,* the generic term that includes phototubes and photosensitive semiconductor devices. Generally speaking, semiconductor devices have supplanted the earlier phototubes.

There are four general types of photocells in common use. One is the *photoemissive* type, which releases electrons when exposed to light. Another is the *photoconductive* type, whose resistance is drastically reduced when illuminated. The *photoconductive-junction* type is similar to the photoconductive type, except that the light is applied to a semiconductor junction. The *photovoltaic* type generates a voltage when exposed to light. In addition to these four basic types, there is the *photoelectromagnetic* photocell, where a semiconductor material is placed in a magnetic field and exposed to light, thus producing a small voltage. The photoelectromagnetic photocell is not in general use. Before going into the details of the four photocell types, let us consider the units of measurement, and the basic terms common to all types.

5-1.1 Units of Light Measurement

Prior to 1948, the *candle power* (cp) was the unit of luminous intensity in common use, and is equal to the luminous flux of one candle when viewed in a horizontal plane. Since 1948, the *candela* (cd) is the official international unit of luminous intensity, and is one-sixtieth of the luminous intensity of 1 sq cm of a blackbody radiator which is at the temperature of solidifying platinum. However, candle power and its related measurements are still in use.

When candle power is used, the *foot-candle* is the unit of illumination of 1 candle per square foot, as well as the illumination at a spherical distance of 1 foot from a one-candle source. Likewise, the *foot-lambert* is the unit of luminance equal to $1/(4 \pi)$ candle per square foot, as well as luminance of a surface emitting or reflecting light at a rate of 1 lumen per square foot.

The *lumen* (lm) is the unit of luminous flux. One lumen is the flux emitted within a solid angle of 1 steradian by a point source having a uniform intensity of 1 candela. The *lux* (lx) is the unit of illumination, and is equal to 1 lumen per square meter. The *candela per square meter* (cd/m²) is the unit of luminance. Luminosity is expressed in *lumens per watt*. The relationship of lumens, footcandles, candelas, and lux is shown in Fig. 5-1. The *angstrom* (A) is the unit of wavelength of light. One A equals 10^{-10} meter (m). The *micron* (u or um) is also a unit of wavelength of light. One micron equals 10^{-6} m (10^{-4} cm). Figure 5-2 shows the relationship between the visible color spectrum and wavelength (expressed in centimeters, angstrom units, and microns).

Light is sometimes measured in *photon energy*, which is expressed in *ergs* or *electron-volts. Noise equivalent power* (NEP) is expressed in *watts* and responsivity in *volts per watt*, when the surface area is given in cm² and the radiation flux in watts/cm².

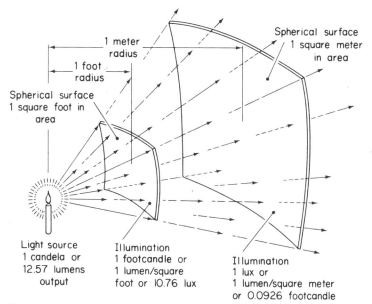

Figure 5-1 Relationship of lumens, footcandles, candelas, and lux (Courtesy General Electric Company).

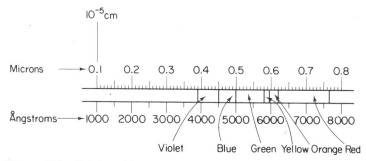

Figure 5-2 Relationship between visible color spectrum and wavelength.

5-1.2 Basic Light Measurement Terms

Light is a form of radiant energy. Strictly defined, only *visible* radiation can be considered as light. *Visible light* is radiant energy which can be detected by the human eye. Visible light wavelengths are between about 0.4 and 0.76 microns as shown in Fig. 5-2. *Infrared light* and *ultraviolet light* are referred to as "radiation" when compared to visible light.

Infrared light is radiant energy in the band of wavelengths between about 0.76 and 100 microns as shown in Fig. 5-3. The portion of the band between 0.76 and 3 microns is sometimes referred to as *near infrared light*.

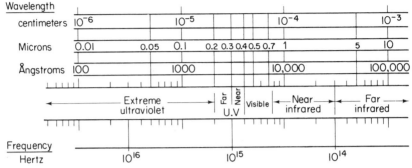

Figure 5-3 The light spectrum.

The portion between 3 and 100 microns is called *far infrared light.* Most of the infrared light band overlaps the *heat radiation* band of electromagnetic radiation. *Ultraviolet light* is radiant energy in the band of wavelengths between about 0.4 and 0.01 microns as shown in Fig. 5-3. Note that X rays (discussed in Sec. 5-2) occupy the adjacent portion of the electromagnetic radiation spectrum, where wavelengths are shorter than 0.01 micron.

Lumiunous flux is the time rate of flow of light (visible light). *Luminous intensity,* in any given direction, is the ratio of the luminous flux emitted by a point source in an infinitesimal solid angle, containing the same given direction, to the solid angle. *Luminance* is the luminous intensity of a surface in a given direction per unit of projected area of the surface as viewed from that direction. Luminance is the photometric (light measuring) quantity equivalent to "brightness". *Illumination* (illuminance) is the luminous flux density on a surface, expressed as luminous flux per unit area (of a uniformly illuminated surface). *Luminosity* is the ratio of the output amplitude to the product of the effective radiation flux density (at a given wavelength) and the detector area (active surface area).

5-1.3 *Photoemissive Light Sensors*

In photoemissive light sensors, electrons are emitted by a cathode when light strikes the cathode. The *phototube,* once a commonly used light-controlled device (such as in alarms), is the most basic of the photoemissive light sensors. The *multiplier phototube* (also known as the photomultiplier tube) is still in use, particularly in one form of radiation counter (as discussed in Sec. 5-3).

Figure 5-4 shows the construction and symbol for a *basic diode-type phototube.* Such a phototube is essentially a vacuum or gas-type diode containing a cathode that has been coated with some photosensitive material, such as cesium, and an anode that normally is positive, relative to the cathode. When light strikes the material coating on the cathode, electrons are emitted and flow to the positive anode. The more intense the light striking the photosensitive cathode, the greater the number of electrons emitted, thus producing an increased electron flow or current. The intensity of the light may be determined by measuring the amount of current produced by the light.

The output current of the basic phototube is very small, typically a few microamperes. If a larger current is needed, the current must be amplified. One method for increasing the current produced by a given amount of light is to use a *multiplier phototube* such as shown in Fig. 5-5; the phototube shown is a linear type. Curved-plate photomultipliers are also used. With either system, the electrons emitted by the cathode are made to strike in succession a series of positive plates, called *dynodes*, before they reach the anode. As each electron strikes a dynode, a number of electrons are knocked off. Thus, a single electron emitted by the cathode will produce several electrons (say five) at dynode 2, etc. If several dynodes are used (typically 10 to

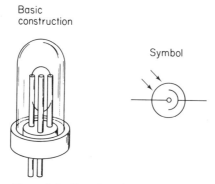

Figure 5-4   Construction and symbol for diode-type phototube.

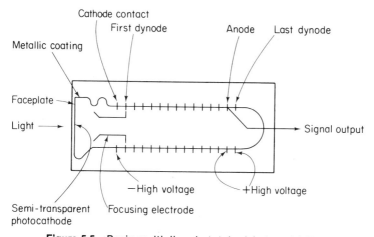

Figure 5-5   Basic multiplier phototube (photomultiplier).

15), several million electrons reach the anode for each electron leaving the cathode. Each dynode in the path between cathode and annode is sufficiently more positive than the previous one to attract electron flow. An anode-cathode voltage difference of 2000 V (or higher) is not uncommon. A typical photomultipler tube will produce about 1 to 2 mA of current flow when exposed to 5 or 10 microlumens (1 microlumen is one-millionth of a lumen).

Figure 5-6 shows the most basic application of a phototube, that of controlling a relay. When the phototube is not exposed to any light, no current flows in the relay circuit. Actually, there may be some small current under these conditions, but it is not sufficient to operate the relay. Such current (the current that flows in total darkness) is known as *dark current*. When light strikes the phototube cathode, current flows and the relay is operated (the normally open contacts close and the normally closed contacts open). Operation and control of relays is discussed further in Chapter 9.

When the relay contacts are arranged so that the circuit is turned on by the presence of light, the circuit is said to be *forward acting*. A *reverse acting* circuit is turned on when light is removed. The relay contacts may be put to any use, such as opening a door, sounding an alarm, etc. Generally, such relay control functions are now performed by solid-state (or semiconductor) photoconductive light sensors described in Secs. 5-1.4 and 5-1.5 or the photovoltaic sensors of Sec. 5-1.6.

5-1.4   Photoconductive Light Sensors

Photoconductive light sensors are semiconductive materials which change their resistance (the resistance is lowered) when exposed to light. For this reason, photoconductive sensors are sometimes referred to as *photoresistive* sensors. Figure 5-7 shows the construction and symbol for some typical photoconductive sensors. As shown, photoconductive material, usually a metal salt, is contained between two conductive terminals to which connecting wires are attached.

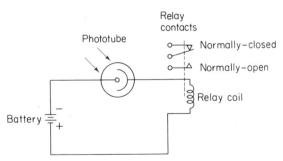

Figure 5-6   Basic application of phototube (relay control).

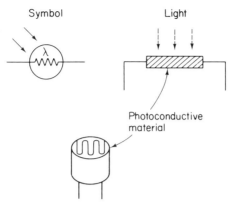

Figure 5-7 Construction and symbol for basic photoconductive light sensor.

Cadmium sulfide (CdS) and cadmium selenide (CdSe) are the most popular photoconductive materials because they are extremely sensitive to light changes, particularly in the visible-light region. When infrared light (specifically) is to be measured, lead sulfide (PbS) and lead selenide (PbSe) are used for the photoconductive material. However, most photoconductive materials are sensitive to all forms of radiation, including X rays and nuclear radiation. Thus, when used to measure visible light only, photoconductive devices must be shielded from possible radiation. The same is true of photovoltaic light sensors described in Sec. 5-1.6.

Any type of photoconductive sensor can be used in the basic control application of Fig. 5-6, as a substitute for the phototube. When light strikes the photoconductive material, the resistance drops from several megohms to a few hundred ohms, or less. When that occurs, the relay is operated by a battery or other external power source. One of the most popular applications for the photoconductive sensor is the photographic exposure meters and for automatic exposure controls in cameras. With these systems, the sensor is used to measure the intensity of light. Since the resistance of the material varies inversely with the intensity of light striking the sensor, the intensity may be determined by measuring the resistance. One method for this measurement is to pass a small constant current through the sensor and then measure the resulting voltage drop on a voltmeter calibrated in units of light intensity. (Or a constant voltage is applied across the sensor and the resulting current flow is measured on a similarly calibrated ammeter.)

There are two problems with photoconductive light sensors. First, an external power source is required in all cases. Second, there is the problem of heat generated when current passes through the sensor material. As is the case with all semiconductors, this heat causes a decrease in resistance (and a further increase in current). This effect can cause damage to the material in extreme cases,as well as possible errors in calibration. Both of these problems are overcome by use of photovoltaic light sensors described in Sec. 5-1.6.

5-1.5 Photoconductive-Junction Light Sensors

Most semiconductor diodes and transistors are photosensitive. That is, current flow increases across the diode and transistor junctions when exposed to light. For this reason, conventional diodes and transistors are sealed from the light by metal cans or other enclosures. When used as light sensors, the diodes (called *photodiodes*) and transistors (*phototransistors*) are constructed so that the junctions are exposed to a maximum of light.

Figure 5-8 shows the construction and symbols for typical photodiodes and phototransistors. Note that the top of the enclosure (which is similar to a typical TO-5 transistor case) is made of a transparent material such as clear plastic. This allows light to be focused on the N-type material in the photodiode or phototransistor and increases current flow in the junctions.

Field-effect transistors (FET) can also be used as phototransistors when they are constructed so that light is focused on the gate as shown in Fig. 5-8. The light causes an increase in gate current, resulting in an increased voltage drop across gate resistance R. This voltage change, in turn, affects the main current flowing from source to drain and produces an increased voltage drop across R_L. This voltage drop is the output signal and can be used to operate a relay or some similar control function.

Both photodiodes and phototransistors have the same deficiencies as other photoconductive light sensors discussed in Sec. 5-1.4. That is, they require external power sources and are subject to the effects of heat when excessive current flows. For these reasons, the photovoltaic light sensors described in Sec. 5-1.6 are more popular.

5-1.6 Photovoltaic Light Sensors

Photovoltaic light sensors are self-generating. That is, they produce a voltage output when exposed to light. Thus, no external power source is required, and the heat produced by the current flow generally causes no problems in practical applications. The *solar cell,* so-called because it is used in space vehicles to convert light energy from the sun to the electrical power required to operate the various instruments in such vehicles, is probably the best-known photovoltaic device. The solar cell can also operate from any other light source and need not be exposed to the sun. Another common application for such sensors is to detect light that passes through holes punched in cards or tapes used in data control devices.

Selenium and silicon are the most popular materials for photovoltaic light sensors (or photocells) when visible light is to be measured. Ger-

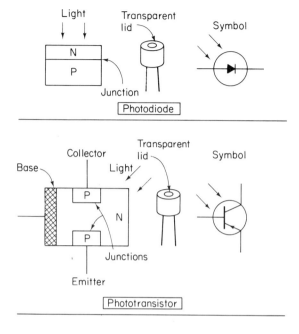

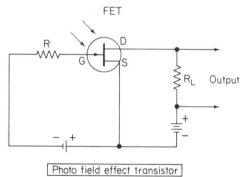

Figure 5-8 Construction and symbols for photoconductive-junction light sensors (photodiodes and phototransistors).

manium can also be used. However, germanium produces maximum output (the most power for a given amount of light) at about 15,500 Å, which is in the infrared region (Fig. 5-3). Selenium and silicon both produce maximum output in the visible light range (about 5700 Å for selenium and 8000 Å for silicon). When infrared light (specifically) is to be measured with a photovoltaic cell, indium-arsenide (InAs) and indium-antimonide (InSb) are used as the photovoltaic material.

Figure 5-9 shows the construction and symbol for a typical silicon

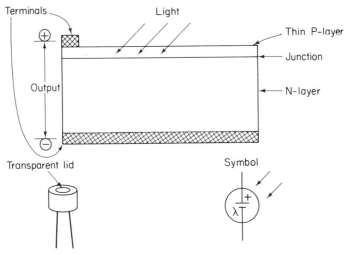

Figure 5-9 Construction and symbol for silicon photovoltaic cell (photovoltaic light sensor).

photovoltaic cell. The cell is essentially a PN junction, with the P layer made so thin that light may pass through to the junction. When light rays strike the silicon atoms near the junction, electrons are agitated and liberated from the atoms. These electrons move about, leaving behind positively charged holes. In the process, a *contact potential* or *space charge* appears across the junction. As a result of electron movement, a current flows, with the N layer acting as the negative terminal of the "battery."

The brighter the light, the more electrons that are set free and the more current that flows in the external circuit connected to the cell terminals. Output current is directly proportional to the total light that falls on the surface of the cell. Usually, photovoltaic cells are rated as to a certain output voltage or power for a given amount of light (specified in foot-candles).

Figure 5-10 shows the construction and symbol for a typical selenium photovoltaic cell. A selenium cell, like a silicon cell, produces electrical current when illuminated. Although selenium is not as efficient as silicon, selenium cells have characteristics more nearly like the human eye. That is, selenium is highly efficient in the range of human vision, but not beyond as is silicon. Therefore, selenium is used in applications where the light resources are similar in level to those that would normally be visible to the human eye.

Selenium photovoltaic cells are made by depositing a thin film or layer of selenium onto a metal base plate. The selenium is then crystalized by heat. Then cadmium oxide is deposited on the selenium to form a junction.

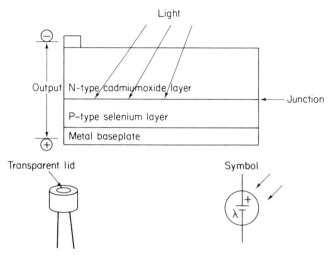

Figure 5-10 Construction and symbol for selenium photovoltaic cell (photovoltaic light sensor).

Note that the selenium layer forms the P region, whereas the cadmium oxide forms the N region.

5-2 X-RAY SENSORS

Because of their penetrating power, X rays are used for many industrial applications, such as detecting the level of liquids in closed tanks, and the thickness of rapidly moving sheets of material such as paper, plastic, or metal. These applications usually require some means for sensing the intensity of the X rays.

One method for sensing X-ray intensity depends on the fact that when X rays strike certain materials, such as calcium tungstate and other substances, a visible glow of light is produced. The more intense the X rays, the greater the glow. The intensity of the glow, in turn, is measured by a photocell of some kind. The output current of the photocell is directly proportional to the glow. Thus, by measuring the photocell output, the intensity of the X rays may be determined. Because the glow produced by calcium tungstate is weak, a multiplier phototube or very sensitive photovoltaic light sensor is generally used. This method of sensing X rays is similar to the scintillation counter method of sensing nuclear radiation discussed in Sec. 5-3.

Another method for sensing X-ray intensity makes use of the fact that X rays affect a cadmium-sulfide crystal in the same manner as visible light (Sec. 5-1.4). That is, the X rays produce a sharp drop in the resistance of the crystal. The more intense the X rays, the lower the crystal resistance. If a steady voltage is applied across the crystal, the current produced varies in proportion to the intensity of the X rays.

5-3 NUCLEAR RADIATION SENSORS

Nuclear or radioactive materials emit nuclear radiations. Such radiations consist essentially of four parts. One part is the *alpha* particle, which is a helium nucleus consisting of two protons and two neutrons. This particle carries a positive charge. The second part is the *beta* particle, which is an electron and carries a negative charge. The third part is the *neutron* particle, which carries no electrical charge. The fourth part is the *gamma ray,* an electromagnetic wave somewhat similar to X rays.

There are two basic methods for converting nuclear radiation into usable electric output signals. Both methods use the reaction of the radiation to a material contained in the sensor or transducer. The first method, called *ionizing transduction,* depends on the production of an *ion pair* in a gaseous or solid material, and the separation of the positive and negative charges to produce a voltage. The second method, called *photoelectric transduction,* uses a *scintillator* material which generates light in the presence of radiation, and a light sensor which provides a voltage output proportional to the light. Before going into the details of the various radiation sensors, let us consider the units of measurement common to all types.

5-3.1   *Units of Radiation Measurement*

The *roentgen* is a unit of X-ray or gamma-ray radiation (exposure dose). One roentgen equals the quantity of radiation whose associated secondary ionizing particles produce ions, in air, carrying one electrostatic unit charge per 0.001293 gram (g) of air. The *rad* is a unit of *absorbed radiation dose* and equals 100 ergs per gram. The *rem* (roentgen equivalent, man) is the unit of RBE (relative biological effectiveness) dose and equals the absorbed dose in rads, times an agreed conventional value of the RBE. The rem was originally defined as the absorbed dose that will produce the same biological effect in human tissue as that produced by 1 roentgen. Typical RBE values are 1 for X rays, 10 for protons, and 20 for alphs particles.

The electron volt (eV) is the unit of radiation energy. One eV equals 1.6×10^{-2} erg. One MeV (mega electron-volt) = 1.6×10^{-6} erg, and is more commonly used for nuclear radiation measurement. The *curie* is a unit of radioactivity, and equals the quantity of any radioactive material in which the number of disintegrations per second is 3.7×10^{10}. *Neutron flux density* is usually expressed in *neutrons per square centimeter-seconds* (n/cm^2 − s). The *window thickness* (of a radiation transducer) is commonly expressed in *milligrams per square centimeter* (mg/cm^2).

5-3.2 Photoelectric Radiation Sensors

Figure 5-11 shows the basic elements of a photoelectric radiation sensor, commonly called a *scintillation counter*. As shown, the elements consist of a scintillator material and a photomultiplier tube (Sec. 5-1.3), both contained in a common housing. A counter and readout may also be included as part of the instrument. The area between the scintillator material and the tube is sealed to visible light, to prevent surrounding light from operating the phototube. The scintillator material, usually crystals such as sodium iodide and zinc sulfide, produce a flash of light, called *scintillation,* each time the crystal is struck by an alpha or beta particle or a gamma ray (or by X rays when certain kinds of crystals are used).

In the scintillation counter of Fig. 5-11, the scintillations of the crystal are picked up by reflecting mirrors and "piped" to the window of the tube by means of a lucite rod. Each particle or ray produces a pulse of anode current at the output of the phototube. The output is applied to a counter circuit such as described in Chapter 8. By counting the number of pulses during a given period of time, the intensity of the radiation may be determined. Note that the scintillation counter is generally more sensitive than the ionization-type radiation sensors described in Sec. 5-3.3.

5-3.3 Ionization-Type Radiation Sensors

Ionization is the process of forming subatomic particles (ions) with a positive or negative charge. Radiation produces ionization in certain gases, crystals, and semiconductors. Thus, if the amount of ionization can be measured, this can be related to the amount of radiation producing the ionization. There are three basic types of radiation sensors that use this principle: the *ionization chamber,* which uses ionization of gas, the *crystal sensor,* which uses ionization in a solid crystal, and the *semiconductor sensor,* which uses ionization in a solid-state semiconductor. The basic elements of all three sensors are shown in Fig. 5-12. In all three instances,

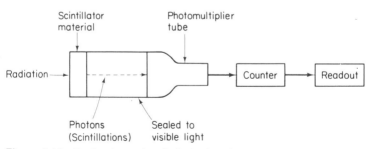

Figure 5-11 Basic elements of photoelectric radiation counter (scintillation counter).

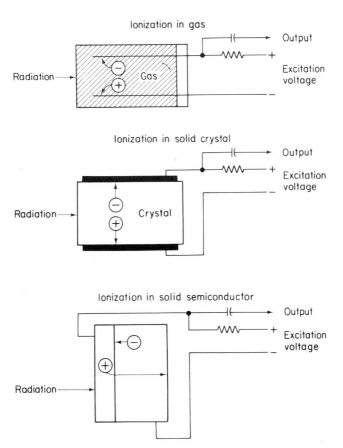

Figure 5-12 Basic elements of ionization-type radiation sensors.

the radiation breaks down part of the material into *ion pairs* with positive and negative charges. The resultant voltage or current is then measured to provide an indication or measure of the radiation.

The crystal shown in Fig. 5-12 has largely been replaced by the semiconductor. One reason for this is that the crystals must be very pure to produce usable outputs. Also, the best known of the ionization-type radiation sensors is the Geiger counter, which uses the Geiger-Muller tube and is a form of ionization chamber. Here, we concentrate on the ionization chamber, Geiger counter, and semiconductor radiation sensors.

Ionization chamber. Figure 5-13 shows the basic elements of an ionization chamber used as a radiation sensor. As shown, the chamber consists of a metal cylinder (the outer electrode) sealed at one end by a thin window. The other end of the cylinder is closed, and a metal rod (the center elec-

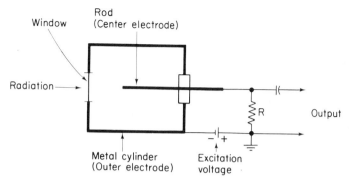

Figure 5-13 Ionization chamber.

trode) protrudes into the cylinder through an insulator. A fixed d-c voltage, sometimes known as the *excitation voltage,* is applied to the electrodes through load resistance R. The output is taken from across R through a coupling capacitor (which is necessary since the excitation voltage is often quite high).

When radiation is applied through the window to the air (or other gas) in the chamber, the air is ionized, and the resulting ions rush to their respective electrodes (positive and negative). In doing so, the ions collide with and ionize other atoms of air, and these new ions join the rush to the electrodes. The entire action is almost instantaneous. In one method of operation, called the *current mode,* the increase in current caused by the flow of charges to the electrodes (the *ionization current*) can be monitored as the average voltage drop across R. In another mode of operation, called the *pulse mode,* the ionization is measured as a single event. The output is then a series of pulses, each pulse generated by the ionization due to one particle.

Generally, the output of an ionization chamber is amplified before it is measured. Different particles cause different amounts of ionization. Thus, the output amplitude can indicate the type of radiation and the average magnitude, if the chamber readout device is so calibrated. Also, although all gases (including air) will produce some ionization, the fill gases used in ionization chambers are selected on the basis of ionization potential (the required excitation voltage) and the energy per ion pair. Typical fill gases include argon, krypton, xenon, hydrogen, neon, helium, nigtrogen, and methane. Sometimes gases are mixed. The gases can be at, above, or below atmospheric pressure but are usually at the surrounding pressure when the window material is very thin.

The types of radiation to which the chamber will respond can be controlled by the thickness and material of the window. Measurement of alpha particles requires very thin windows, often of mica or nylon. Beta particles, as well as gamma rays and X rays, can be measured with thin metal windows. Sometimes, the window is omitted and the chamber walls are made sufficiently thin for radiation to pass.

Geiger counter.   Figure 5-14 shows the basic elements of a Geiger counter, which is probably the best known of the radiation sensors. The heart of the Geiger counter is the Geiger-Muller tube, which is a form of ionization chamber. The tube is essentially a diode, consisting of a cathode in the form of a long metal cylinder, and an anode in the form of a fine wire running through the center of the cylinder. Both are mounted in a thin-walled, air-tight, glass envelope sealed by an extremely thin window through which the radiations may pass. The air is evacuated from the envelope, and a small amount of some inert gas such as argon is added.

The anode is made positive and the cathode negative by means of a battery whose voltage is *just below the ionization point of the gas* in the envelope. When radiation penetrates the window, some of the gas atoms are ionized. The resulting negative ions rush to the positive anode, and the positive ions rush to the negative cathode. During their progress the ions collide with some of the gas atoms and produce more ionization, thus producing a pulse of current through the tube and load resistor R. The resulting voltage drop across R is the output voltage.

It might be expected that ionization will continue once established. However, note that resistor R (which has a high resistance) is in series with the anode of the tube and the battery. As long as the gas is not ionized (and no current flows through the tube) there is no voltage drop across resistor R, and the voltage on the anode of the tube is the same as that of the battery. When the gas ionizes and current flow through the tube and resistor R, a large voltage drop takes place across R. That voltage drop is sufficient to reduce the voltage on the anode *below the ionizing point of the gas.* The gas deionizes and current stops flowing until radiations again penetrate the tube envelope.

In this manner, a series of alpha or beta particles or bursts of gamma rays cause a series of current pulses to flow through the tube and the anode

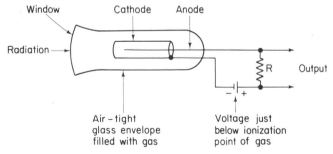

Figure 5-14   Geiger-Muller tube and basic circuit (Geiger counter).

circuit. The output pulses from the tube are amplified and registered on some indicating devices as flashes of light or as clicks from a loudspeaker. By counting the number of flashes or clicks, it is possible to tell how many particles entered the tube during a given period of time, and thus obtain an indication of radiation intensity. In some Geiger counters, the pulses are stored up and the cumulative result shown on a meter calibrated in units of radioactivity (roentgen, rem, rad), one pulse for each alpha or beta particle or gamma ray.

Semiconductor radiation sensors. The photoconductive and photoconductive-junction light sensors described in Secs. 5-1.4 and 5-1.5, respectively, are also used as radiation sensors. The radiation has the same net effect on the photoconductive materials and the junctions of semiconductors as does light. That is, the current flow is increased when either light or radiation is applied to the material. For example, the cadmium-sulfide photoconductive sensors described in Sec. 5-1.4 have been used for some types of radiation measurement, particularly X rays. When so used, the cadmium sulfide is enclosed in a metal housing to prevent surrounding light from affecting the material. The housing is provided with a thin beryllium radiation window to permit the X rays to pass to the material. A typical circuit is shown in Fig. 5-15. When an excitation voltage is applied, current flows through load resistor R and the cadmium sulfide. When X rays are applied, the resistance of the cadmium sulfide drops, increasing the voltage across R and producing a voltage pulse which is used as the output.

The junction-type semiconductor radiation sensors are found in two forms: intrinsic (or natural) crystals and *extrinsic,* which are manufactured to provide the desired junction characteristics. Except in rare cases, the extrinsic sensors are used for all applications, since they are more sensitive and their characteristics can be controlled.

Extrinsic semiconductor sensors are of two basic types: the *surface barrier type* and the *diffused junction type,* both shown in Fig. 5-16. The sur-

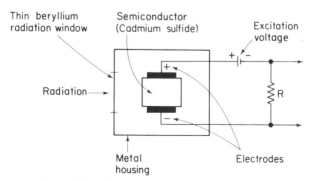

Figure 5-15   Semiconductor radiation sensor.

face barrier type usually consists of N-type single-crystal silicon on one surface of which a P-type layer of silicon dioxide is formed. This thin layer is covered by a very thin film of vacuum-evaporated gold. The diffused junction type is commonly made by a very shallow diffusion of N-type material into a base of P-type single-crystal material (usually silicon).

With either type, there is a *depletion layer* or region (also known as the *space charge region*) close to the junction. It is in this depletion region that electron-hole pairs are produced by radiation in the same manner in which ion pairs are produced in the gas of gas-filled (ionization-type) radiation sensors. The depth of the depletion region (the *depletion depth*) varies with the resistivity of the N- or P-type single-crystal material and with the reverse-bias voltage applied across the two terminals as shown in Fig. 5-16. For some applications, the depletion region depth is increased by an intrinsic semiconductor between the N- and P-type materials. These devices are known as *PIN junction radiation sensors.*

5-3.4 · Neutron Radiation Sensors

Because they carry no electrical charge, neutrons are very difficult to detect. In some cases, neutrons can produce occasional ionizing particles by transferring some of their energy to other particles with which they collide. However, in practical applications neutrons are sensed and measured by indirect means. When neutrons strike the atom of a certain uranium isotope, the uranium atom splits into two or more parts. This splitting is known as

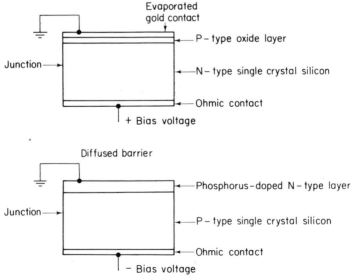

Figure 5-16   Extrinsic semiconductor sensors.

fission and results in the release of radiation (alpha and beta particles and gamma rays) from the atom. Such radiation can be detected and measured as described. Thus, by measuring the intensity of the radiation, the intensity of the neutrons producing the radiation may be determined.

There are several different uranium isotopes, of which the atom with a mass number of 238 is the most common. But only the uranium atom with a mass number of 235 produces fission when struck by neutrons. A simple neutron sensor consists of a semiconductor junction diode, the top surface of which is coated with a very thin layer of uranium 235 as shown in Fig. 5-17. The junction diode is reverse-biased so that only a small reverse-current leakage flows across the junction. When the junction is struck by radiations resulting from the fission (caused by neutrons striking atoms in the uranium 235 layer), the reverse current is increased. By measuring the increase in current, the intensity of the neutrons may be determined. Only slow neutrons produce fission of uranium atoms. Where fast neutrons are to be detected, a thin layer of paraffin converts the fast neutrons to slow neutrons.

Another similar method for detecting neutrons consists of coating the junction diode with a thin layer of some chemical containing lithium atoms. When an atom of lithium is struck by a neutron, the atom breaks down into two charged particles. One is an alpha particle (helium nucleus) and the other is a *triton* (the nucleus of a rare type of hydrogen atom with a mass number of 3). These charged particles affect the diode like other types of radiation. That is, the particles produce an increase in current flow across a junction.

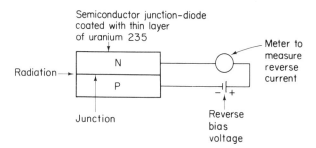

Figure 5-17 Neutron radiation sensor.

6

TEMPERATURE SENSORS

Temperature sensors are generally classified by the method of sensing. In this chapter, we cover the six most commonly used methods for sensing temperature: *bimetallic, fluid-pressure, resistive, thermocouple, radiation,* and *oscillating crystal.*

6-1 TEMPERATURE-SENSING BASICS

Operation of the bimetallic sensor (Sec. 6-2) is based on the uneven rate of expansion and contraction when dissimilar metals are welded together and heated or cooled. The fluid-pressure sensor (Sec. 6-3) responds to expansion and contraction of fluids (or gases) in a sealed chamber. Operation of resistive sensors (Sec. 6-4) is based on the change of resistance in metals and semiconductors when heated or cooled. The thermocouple sensor (Sec. 6-5) responds to the voltage produced when two dissimilar metals are joined and heated. Operation of the radiation sensor (Sec. 6-6) is based on the radiation produced when a substance is heated to an extreme. The oscillating crystal (Sec. 6-7) responds to variation in output frequency when the temperature of a quartz crystal is changed. Before going into the details of specific temperature-sensing methods, let us consider the units of measurement and the basic terms common to all types.

6-1.1 Units of Temperature Measurement

Temperature is expressed in *degrees Celsius* (°C), *degrees Fahrenheit* (°F), *degrees Kelvin* (°K), or *degrees Rankine* (°R), depending on the temperature scale used. Degree Kelvin is the official international unit and is also known as the *thermodynamic scale.* Kelvin is defined by assigning the temperature value 273.16 °K to the *triple point of water* (Sec. 6-1.2). Basic temperature scale conversion is as follows:

$$°C = (°F - 32) \times \frac{5}{9}, \quad °F = \frac{9}{5} \times °C + 32,$$

$$°K = °C + 273.15 \quad °R = °F + 459.67$$

	CELSIUS	FAHRENHEIT	KELVIN	RANKINE
Steam point:	100 °C	212 °F	373.15 °K	671.67 °R
Ice point:	0 °C	32 °F	273.15 °K	491.67 °R
Absolute zero:	− 273.15 °C	− 459.67 °F	0 °K	0 °R

6-1.2 Basic Temperature Measurement Terms

The *temperature* of a body is its thermal state considered in reference to its power of communicating heat to other bodies. *Heat* is energy transfer, due to temperature differences, between a system and its surroundings or between two systems, substances, or bodies.

Heat transfer is the transfer of heat energy by one or more of the following methods:

Radiation by electromagnetic waves; *conduction* by diffusion through solid material or through stagnant fluids (liquids or gases); *convection* by movement of a fluid (including air) between two points.

The *ice point* (273.15 °K) is the temperature at which ice is in equilibrium with air-saturated water at a pressure of 1 atmosphere (Sec. 3-2.1). The *steam point* (100 °C) is the temperature at which steam is in equilibrium with pure water at a pressure of 1 atmosphere. The *triple point of water* (273.16 °K) is the temperature at which the solid, liquid, and vapor states of water are all in equilibrium. Notice that it is just above the ice point.

Thermal equilibrium is a condition of a system and its surroundings (or two or more systems, substances, or bodies) when no temperature difference exists between them (no more heat transfer occurs between them).

6-2 BIMETALLIC TEMPERATURE SENSORS

One of the most common bimetallic temperature sensors is the *thermostat* shown in Fig. 6-1. The heart of this device is a bimetallic strip consisting of two dissimilar metals welded together. Each metal has a different rate of ex-

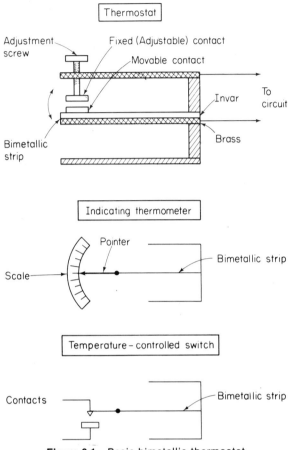

Figure 6-1 Basic bimetallic thermostat.

pansion due to heat. Metals commonly used for the strips are brass, which has a relatively large rate of expansion, and invar, an alloy of nickel and iron, which has a relatively small rate of expansion. When the bimetallic strip is heated, the greater expansion of the brass causes the free end of the strip to bend upward as shown. When cooled, the strip returns to its normal position (straight). The amount the bimetallic strip bends is directly proportional to the degree of heat applied.

6-2.1 *Thermostat-Type Indicating Thermometer*

The basic thermostat of Fig. 6-1 can be used as an *indicating thermometer* by fastening a pointer to the free end of the strip and allowing the pointer to move over a calibrated scale. The basic thermostat can also be used with a motion transducer (Chapter 2) to produce a signal to actuate some control device. Likewise, a thermostat can be used as a *temperature-*

controlled switch by fastening a contact point at the free end of the bimetallic strip so that, when the heat causes the strip to bend a predetermined amount, the movable contact touches a fixed contact, thus closing the control circuit. When the temperature of the strip decreases, the strip tends to straighten, thus separating the contact point and opening the circuit. The temperature at which the contact point will touch can be controlled, within a certain range, by means of the adjustment screw, which brings the fixed contact nearer to or farther away from the movable contact. In this way, the amount of the strip must bend before the circuit is closed can be controlled.

6-2.2   Circuit Breaker

A thermostat can also be used as an automatic *circuit breaker,* used as protection against excessive current flow in an electrical circuit. Generally, circuit breakers are connected in series with the circuit to be protected. Normally, the contact points touch and the circuit is completed as shown in Fig. 6-2. As current flows through the strip, the resistance of the strip produces a certain amount of heat. If the current is normal (at or below the rating of the circuit breaker) the heat is not great enough to separate the contact points. Should the current rise above the rating, the heat becomes great enough to cause the strip to bend sufficiently to separate the contact points. The circuit is then opened, and the current flow stops.

6-2.3   Disk Switch

Figure 6-3 shows a version of a thermostat, called the *bimetallic disk switch,* used to prevent overheating in an electric motor. The switch is mounted inside the motor, and is connected in series with the motor and power line. Note that current passes through a heater mounted near the strip. When the current rises above the switch rating, the heat produced by the heater is sufficient to cause the disk to change its curvature, thus opening the circuit and stopping the flow of current. This same condition can be produced by heat from the motor. When the motor overheats to a

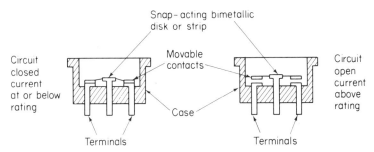

Figure 6-2   Bimetallic thermostat used as a circuit breaker.

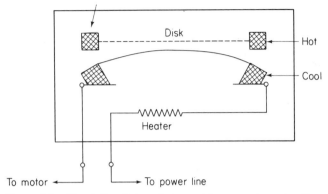

Figure 6-3 Bimetallic disk switch used to prevent overheating in an electric motor.

dangerous level, the heat opens the switch contacts and removes power to the motor. After the motor stops and cools sufficiently, the contact again closes, current is applied, and the motor starts to run. Thus, the motor is protected from overheating, even if the current is not above a desired level.

6-2.4 Bimetallic Thermal Valve

The principle of the bimetallic disk can also be applied to a *thermal valve*. With such an arrangement, the disk transmits its force mechanically to open or close an opening, valve, etc., rather than to make or break an electrical circuit. Note that, because of the springiness of the disk, the action of the disk switch is not gradual. The disk snaps from one curvature to another when the surrounding temperature reaches a certain predetermined level. When the temperature falls to another predetermined level, the disk snaps back to its original position. The snapping action can be transmitted mechanically.

6-2.5 Bimetallic Thermometer

Bimetallic temperature sensors need not be limited to provide only on-off control. For example, a *bimetallic thermometer* can be produced by using the same principle as shown in Fig. 6-4. Such a thermometer consists of a long bimetallic strip formed in the shape of a flat spiral. One end of the spiral is fixed, with the outer end free. A pointer, attached to the free end, moves over a calibrated scale. As the bimetallic spiral is heated, the unequal expansion of metals causes the spiral to tighten, moving the pointer toward the high end of the scale. When the spiral is cooled, it unwinds and moves the pointer toward the low end of the scale. The tightening and unwinding of the spiral is proportional to temperature, as read on the scale. If desired, a motion transducer (Chapter 2) can be attached to the free end of the

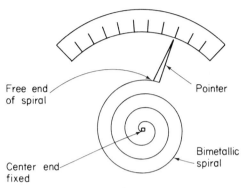

Calibrated scale (°C or °F)

Free end of spiral

Pointer

Center end fixed

Bimetallic spiral

Figure 6-4 Bimetallic thermometer.

spiral, thus producing an output signal proportional to the instantaneous value of the temperature.

6-3 FLUID-PRESSURE TEMPERATURE SENSORS

When a fluid (liquid or gas) is heated, the fluid expands in direct proportion to the heat applied. Conversely, if a fluid is cooled, it contracts in direct proportion to the decrease in heat. This principle can be used to provide a temperature sensor. The most common temperature sensor using the fluid-pressure principle is the *mercury thermometer*. Typical mercury thermometers have mercury enclosed in a bulb of metal or glass. As the bulb is heated, the mercury tends to expand. However, since the volume of the bulb is fixed, the internal pressure is increased, and the increase is directly proportional to the increase in temperature. When the temperature decreases, the pressure also decreases in direct proportion. The bulb is attached to a tube which is partially filled with a column of mercury. The mercury settles into the bulb and mixes with the mercury in the bulb. When the temperature increases, internal pressure increases, and the mercury column rises. The corresponding temperature is read out on a scale next to the mercury column.

Mercury is not always used in fluid-pressure temperature sensors. In some systems known as *vapor-filled* sensors, the bulb is partially filled with a volatile liquid which, when heated, vaporizes and fills the remainder of the bulb and tube. Other systems are *gas-filled,* where the bulb and tube are filled with an inert gas. Both the gas-filled and vapor-filled systems are generally used with systems where the fluid-sensor operates in conjunction with a secondary transducer to indicate the temperature. The basic form of such a system is shown in Fig. 6-5, where the bulb and tube are connected to

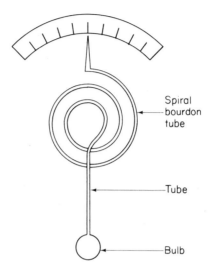

Calibrated scale

Spiral bourdon tube

Tube

Bulb

Figure 6-5 Fluid-pressure (filled bulb) temperature sensor using spiral Bourdon tube.

a Bourdon tube (Sec. 3-2.4), which translates pressure changes into motion that moves a pointer over a scale calibrated in degrees of temperature.

6-4 RESISTIVE TEMPERATURE SENSORS

There are two basic types of resistive temperature sensors: the *conductive type* and the *semiconductor type*. Both types operate on the principle that the resistance of conductors and semiconductors changes with temperature.

6-4.1 Conductive-Type Temperature Sensors

Generally, the resistance of a metal increases as temperature rises. Thus, an indication of temperature may be obtained by measuring resistance. The most common type of resistive temperature sensor is shown in Fig. 6-6. The temperature sensing element is a coil of fine wire such as copper, nickel, or platinum. All these metals have a *positive temperature coefficient* (they increase in resistance with increases in temperature). Copper was used extensively in the past but has been replaced by nickel for most applications. Platinum is used where very precise measurements are required and where high temperatures are involved. Typically, nickel is used in the $-100°$ to $+300°C$ range. Platinum is used in to -265 to $+1100°C$ range. The metals used for the wire-sensing element in conductive-type temperature sensors

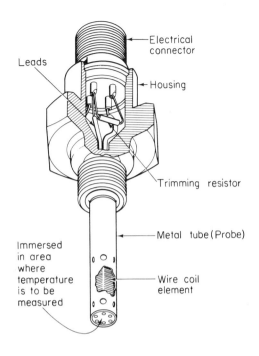

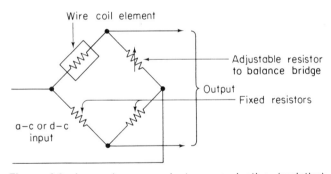

Figure 6-6 Immersion or probe-type conductive (resistive) temperature sensor.

are rated as to temperature coefficient and resistivity. *Temperature coefficient* describes the amount of resistance change for a given temperature change and is usually expressed in ohms per °C. Resistivity describes the resistance of a metal wire (of given length and size) at a given temperature. Generally, the highest resistivity and temperature coefficient is the most desirable, although in some applications a linear temperature coefficient is of greater importance.

The temperature sensors such as shown in Fig. 6-6 are known as *immersion* or *probe* types, since they are designed to measure the temperature within inaccessible areas. The wire coil element is usually enclosed in a metal tube for protection and is used as a probe to be inserted into the area where the temperature is to be sensed. A set of leads running through the tube connects the coil to the rest of the circuit.

Other forms of resistive temperature sensors are shown in Fig. 6-7. These types are used to measure *surface temperatures*. One surface temperature sensor is similar to a bonded-wire strain gage (Chapter 2). The other form, shown in Fig. 6-7, has a coiled-wire sensing element bonded to the inside of a small case at the bottom portion of the coil. Insulating material fills the inside of the case and prevents coil motion. Both types of surface temperature sensors can be cemented, welded, or bolted to a surface.

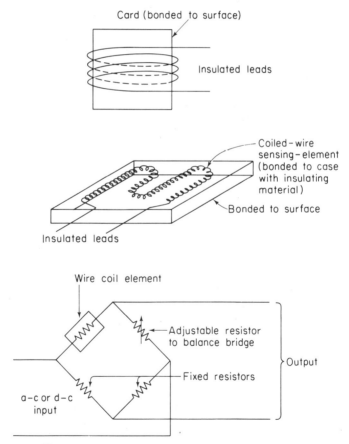

Figure 6-7 Resistive surface temperature sensors.

In addition to the wire-type resistive temperature sensors, *metal-film element* sensors are being used for many applications. Such sensors use a metal film deposited on an insulator, instead of a wire element wound in coil form. A typical metal film element consists of a platinum film less than 1 microinch thick deposited on a ceramic disk or probe. Even though platinum is used, the cost of metal film elements is generally less than wire elements and far more rugged. A typical temperature range for a platinum metal-film element sensor is $-430°$ to $+800°C$.

Resistive temperature sensors are generally used as part of a bridge circuit. That is, the resistance element of the sensor forms one leg of the bridge as shown in Figs. 6-6 and 6-7. The remaining legs of the bridge are formed by fixed resistors. In some instances, one leg of the bridge is formed by an adjustable resistor, so that the bridge can be balanced at a given temperature.

6-4.2 Semiconductor Temperature Sensors

Generally, the resistance of a semiconductor decreases as temperature rises. Thus, semiconductor temperature sensors have a *negative temperature coefficient*. The most common type of semiconductor temperature sensor is the *thermistor* shown in Fig. 6-8. Although thermistors come in various

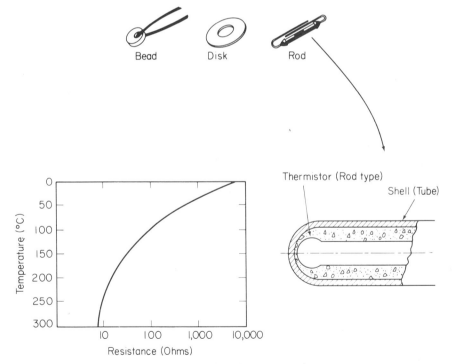

Figure 6-8   Thermistor semiconductor temperature sensor.

sizes and shapes as shown, they all contain a semiconductor of ceramic material made by sintering mixtures of metallic oxides such as manganese, nickel, cobalt, copper, iron, and uranium, and formed into small glass-enclosed beads, disks, or rods. Because of their small size, thermistors are used where other temperature sensors cannot be used.

A typical resistance-temperature curve for a thermistor is shown in Fig. 6-8. Note that the curve is not linear. Nonlinearity of the temperature coefficient is one of the deficiencies of thermistors. Others are the low currents with which they operate (usually less than $100 \mu A$), and their tendency to drift (in temperature) over long periods of time. However, thermistors do have a fast temperature response time. That is, they change resistance quickly (compared to the resistance element of Sec. 6-4.1) in response to temperature changes. Thermistors are generally used in the temperature range of $-75°$ to $+250°C$. Special thermistors have been developed for temperature measurements in the range below $24°K$.

In addition to thermistors, semiconductor temperature sensors are made from germanium and silicon crystals, carbon resistors, and gallium-arsenide diodes.

Germanium crystal sensors have been developed primarily for temperature measurements in the $1°$ to $35°K$ regions. The resistance-temperature characteristics of germanium crystals show a nonlinearity similar to that of thermistors. Germanium crystal transducers include coated crystals, crystals encapsulated and hermetically sealed in metal covers, and immersion probes.

Silicon crystal sensors are generally designed as small, very thin wafers with leads at opposite ends, primarily for surface applications. Their resistance-temperature curve is similar to that of metal wire between about $550°F$ and a point lying between $0°F$ and $-100°F$. Below that point, the curve resembles that of a thermistor. Thus, silicon crystal sensors differ from thermistors and germanium elements in that their temperature coefficient is positive above $-50°C$ and their resistance-temperature characteristics are much more linear, particularly over their most usable range, $-50°C$ to $+250°C$. Below $-50°C$ their temperature coefficient is negative, and the slope of their resistance-temperature characteristics increases sharply.

Commercially available *carbon resistors* have been used as temperature-sensing elements in the region below $60°K$, where their behavior is similar to that of semiconductor temperature sensors. Carbon resistors have a high resistivity and negative temperature coefficient. This assures a relatively high resistance in the lower portion of their range.

Gallium-arsenide diodes (PN junctions) have been developed for cryogenic and low-temperature measurements. When a constant forward current is maintained through a gallium-arsenide diode, the forward voltage varies almost linearly with temperature over the range 2° to 70 °K and 100° to 300 °K, although the slope of the voltage-temperature curve at constant current is not the same for the two ranges.

6-5 THERMOCOUPLE TEMPERATURE SENSORS

Thermocouple temperature sensors are also known as *thermoelectric temperature transducers*. The basic thermocouple circuit shown in Fig. 6-9 consists of a pair of wires of different metals joined together at one end (the sensing junction) and terminated at their other end by terminals (the reference junction) and maintained at an equal and known temperature (reference temperature). The circuit is completed by a load, generally composed of a signal conditioning circuit. When there is a temperature difference between the sensing and reference junctions, a voltage is produced and flows through the load. This characteristic is known as the *thermoelectric effect*.

The amount of voltage produced by thermoelectric effect depends on the metals used and the temperature difference between the two junctions. Figure 6-10 shows the voltage generated by some typical metals over a range of temperatures. There are similar graphs and elaborate charts available for a wide variety of metals. The most commonly use thermocouples are Chromel-Alumel, iron-constantan, copper-constantan, Chromel-constantan, and platinum-platinum/rhodium alloy. Chromel (90% nickel, 10% chromium) and Alumel (95% nickel, 5% alaminum-silicon-manganese) are registered trade names of Hoskins Mfg. Co., Detroit, Michigan. The iron-constantan combination is generally used for temperatures up to about 1600 °F, and the Chromel-Alumel combination for temperatures up to about 2100 °F. The platinum-platinum/rhodium alloy can be used for even higher temperatures.

The thermocouple generally is encased in a metal tube for protection as

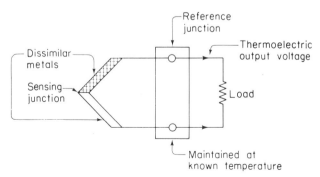

Figure 6-9   Basic thermocouple circuit.

shown in Fig. 6-11. The sensing junction is placed within the device whose temperature is to be sensed. The voltage developed across the reference junction and load is measured and compared with a chart or graph to find the temperature at the sensing junction. In some cases, the small voltage produced by the thermocouple is amplified to operate a control device.

A *thermopile* is a combination of several thermocouples of the same materials connected in series as shown in Fig. 6-12, which is a schematic diagram of a typical Chromel-Alumel thermopile. The output of a thermopile is equal to the output from each thermocouple multiplied by the number of thermocouples in the thermopile. All reference junctions must be at the same temperature. The polarities of voltage shown in Fig. 6-12 are

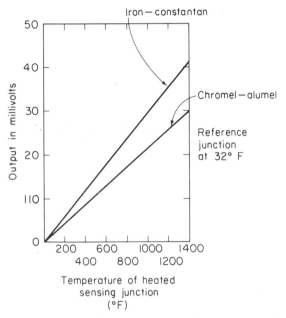

Figure 6-10 Voltage developed by some typical metals connected as a thermocouple.

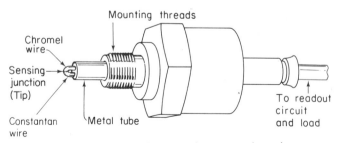

Figure 6-11 Immersion-type thermocouple probe.

based on a temperature higher at the sensing junctions than at the reference junctions.

Foil thermocouples such as shown in Fig. 6-13 are available for surface temperature measurements where thin, flat sensing junctions are required. Each half of the symmetrical foil pattern is made from a different thermocouple material. A butt-type junction is formed at the point where the two foil segments meet. There are two basic types of foil thermocouples: a free-foil style with removable base and a matrix type with foil imbedded in a very thin plastic material. Either type can be bonded to flat or curved conducting or nonconducting surfaces.

6-6 RADIATION PYROMETER

Where very high temperatures such as those of metals being melted in an electric furnace are to be measured, most of the temperature sensors described thus far cannot be used because they would be destroyed by high heat. The radiation pyrometer, shown in Fig. 6-14, overcomes that problem. This pyrometer consists of a metal housing containing a fused-silica lens and a thermopile. The lens end is inserted into a small opening in the side of the furnace. Radiation from the heated material enters the lens and is focused on the thermopile. A fused-silica lens is used because it can withstand high temperatures. A cable from the other end of the pyrometer carries the generated voltage from the thermopile to the signal conditioning or measuring circuit. In some pyrometers, the thermopile is replaced by a phototube or photocell (Chapter 5). Both the phototube and photocell re-

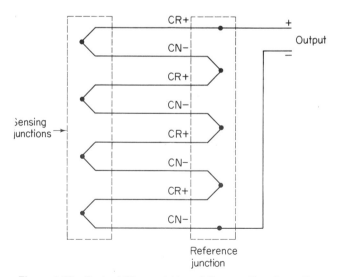

Figure 6-12 Typical Chromel-Alumel thermopile schematic.

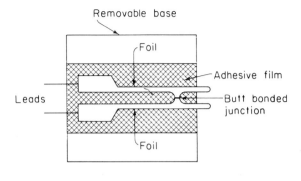

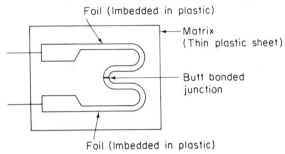

Figure 6-13 Typical foil thermocouples.

spond to radiant energy and produce an output current in proportion to that energy (which provides a measure of temperature).

6-7 OSCILLATING-CRYSTAL TEMPERATURE SENSORS

The quartz crystal used to control the frequency of oscillator circuits is temperature sensitive. That is, the frequency of oscillation is changed when the temperature of the crystal is changed. This condition can be used to measure temperatures in a range from about $-40°$ to $+230°C$. When used to control the frequency of an oscillator (operated at about 28 MHz), the quartz crystal shows a frequency versus temperature slope of about 1 kHz/°C. The output of the oscillator is mixed with that of a reference oscillator so that a mixed (or "beat") frequency is obtained. This mixed frequency is displayed on a frequency counter.

A typical quartz crystal temperature sensor system is shown in Fig. 6-15. The crystal sensing element (a thin quartz crystal wafer) is contained in the tip of a probe connected by cable to the electronics unit (which contains a power supply, oscillators, signal conditioning, and display equipment). The frequency output obtained by mixing the transducer-oscillator signal with a reference-oscillator signal is converted into a number of counts per unit time (by a counter such as discussed in Chapter 8) and indicated on a digital display device (such as discussed in Chapter 10) either in °F or °C.

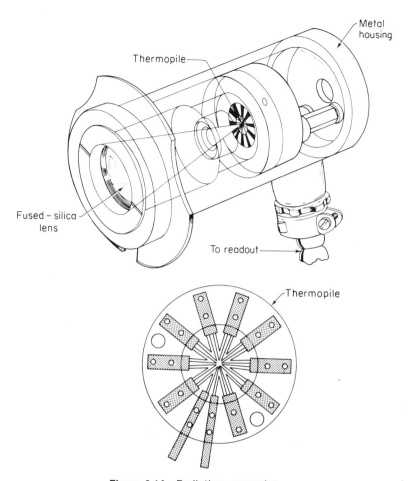

Figure 6-14 Radiation pyrometer.

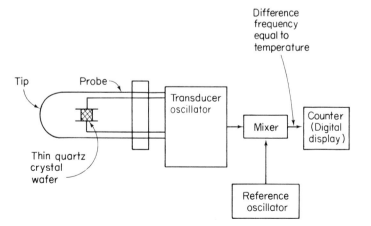

Figure 6-15 Quartz crystal temperature sensor system.

7

SUMMARY
OF SENSING METHODS

In this chapter, we summarize all the sensing techniques discussed in Chapters 1 through 6. We also describe the basics of signal conditioning. Then we show how these basic techniques can be used to sense or measure certain specific quantities such as thickness, proximity, density, chemical content, and sound.

7-1 BASIC MEASURING AND CONTROL SYSTEMS

As shown in Fig. 7-1, a basic measuring system consists of a *transducer,* which converts the quantity or condition to be measured (the *measurand*) into a usable electrical signal. Note that the terms *transducer* and *sensor* are often interchanged, both in this book and in the control/instrumentation field. Specifically, a transducer is the *complete device* used to provide an output in response to a specific measurand. A sensor is the element in a transducer that actually senses the measurand. Thus, a transducer may only contain a sensor or more likely, will also include a transducing element and possibly signal conditioning circuits.

Sometimes a transducer signal output is usable "as is." Most often, however, a transducer signal is modified by signal conditioning circuits. For a simple measuring system such as shown in Fig. 7-1 the conditioned signal is then applied to a display, such as an analog or digital meter, chart recorder, or numerical printout, where the measurand appears in a readable form such as numbers, degrees, etc.

For a control system such as shown in Fig. 7-2 the output of the signal conditioner is applied to a *controller* or other control device. In turn, the controller produces an output (based on the measured input) to operate control devices such as valves, actuators, motors, etc. For example, in a very simple control system, a transducer measures pressure in a tank and produces a corresponding output that is applied to a controller through signal conditioning circuits. The controller then operates a valve which controls the pressure in a tank. With such a closed loop system (Chapter 1), tank pressure can be maintained at any desired value within certain limits.

Signal conditioning circuits can be part of a transducer or a controller, or ing of a bridge-type strain-gage transducer. The upward arrows on the signal conditioning circuits, and two basic reasons for such variety. First, a controller may be a simple electromechanical device such as a relay or may be a very complex unit such as a digital computer (possibly a microcomputer based on operation of a microprocessor). The signals required by each of these controllers are quite different. A relay can be operated with simple d-c voltage from a resistance-type transducer, but a computer requires a digital pulse input.

Second, even if only one controller is used, there are many types of transducers that must be reconciled to a common input. For example,

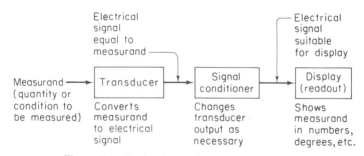

Figure 7-1 Basic electronic measuring system.

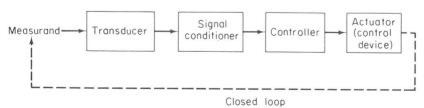

Figure 7-2 Basic electronic control system.

assume that a controller requires a d-c input and that some of the transducers in the control system produce an a-c output. Under these conditions, the a-c outputs must be *rectified* and *demodulated* to direct current. Or, assume that a controller requires signals in the 5 V range and some of the transducers produce outputs in the millivolt range. Obviously, these signals must be *amplified* to the 5 V range.

7-2 BASIC TRANSDUCER DEFINITIONS AND METHODS

In this section, we describe how a transducer can be classified or described and summarize all the transduction methods in common use.

7-2.1 Describing a Transducer

There are many ways to describe a transducer. One way is to answer the following six questions:

1. What is the *measurand,* or what is to be measured by the transducer (acceleration, motion, etc.)?

2. What is the *transduction principle* or the nature of operation (resistive, capacitive, etc.)?

3. What is the *sensing element* (bellows, moving arm, etc.)?

4. What is the *measuring range,* or upper and lower limits of the measurand to be measured (± 5 *g*'s of gravity, $\pm 50°$ of angular rotation, etc.)?

5. What is the *output signal range,* or upper and lower limits of the output signal (± 5 V d-c, 0 to 10 mV, etc.)?

6. What are the *special features,* if any (signal conditioning circuits as part of the transducer, waterproof case, etc.)?

7-2.2 Summary of Transduction Methods

The following is a summary of the transduction or sensing methods discussed in Chapters 1 through 6.

Resistive transduction. As shown in Fig. 7-3, the measurand is converted into a change of resistance in a resistive transducer. The resistance change can be in either a conductor or a semiconductor and can be accomplished by various means. The most common means of changing resistance is by sliding a wiper or contact arm along a resistance element. Other ways to change resistance include drying or wetting of materials such as salts, applying mechanical stress, and heating or cooling a resistance element.

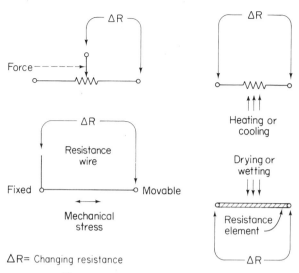

$\Delta R=$ Changing resistance

Figure 7-3 Resistive transduction.

Strain-gage transduction. The strain gage is a special form of resistive transducer, as shown in Fig. 7-4. Here, the measurand is converted into resistance change by mechanical stress or strain. Typically, a strain-gage transducer is used in a bridge circuit, a fixed reference or excitation voltage is applied, and the output is a variable voltage the amplitude of which indicates the amount of strain. Either a-c or d-c excitation voltages can be used, depending on what is required at the output (to match other transducers in the system).

Figure 7-4 shows both the physical construction and the schematic wiring of a bridge-type strain-gage transducer. The upward arrows on the schematic indicate increasing resistances, whereas downward arrows show decreasing resistance. Note that resistances A and D increase but resistances B and C decrease for a strain in one direction. If the direction of strain is reversed, the resistance changes reverse (A and D decrease, B and C increase. These resistance changes cause the output voltage to change polarity. Thus, output voltage indicates both direction and amount of strain (as shown by the polarity and amplitude of the voltage, respectively).

Potentiometric transduction. As shown in Fig. 7-5, potentiometric transduction is another form of resistive transducer. Here, a resistance element is connected as a potentiometer, rather than as the simple rheostat shown in Fig. 7-4 for resistive transduction. In the circuit of Fig. 7-5, movement of the wiper arm causes a change in the ratio (*resistance ratio*) between the resistance from one element end to the wiper arm and the total element resistance. For a typical application, an a-c or d-c reference voltage is ap-

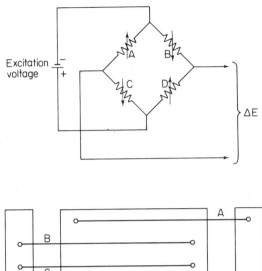

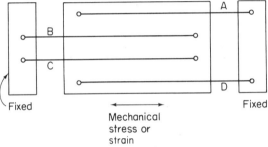

Figure 7-4 Strain-gage transduction.

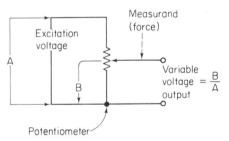

Figure 7-5 Potentiometric transduction.

plied, and the output is a *voltage ratio*. For example, if the reference voltage is 10 V and the arm is at the half-way (50%) point on a resistance element, the output will by 5 V. If the arm is moved up to the 75% point by the measurand, the output will be a 7.5 V.

Capacitive transduction. As shown in Fig. 7-6, the measurand is converted into a change of capacitance in a capacitive transducer. A capacitor consists essentially of two conductors or plates separated by an insulator or dielectric. In a capacitive transducer, the change of capacitance occurs typically when a displacement of a sensing element causes one plate to move

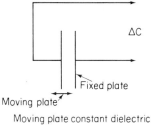

Fixed plate

Moving plate

Moving plate constant dielectric

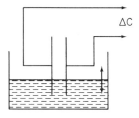

Fixed plates changing dielectric

Figure 7-6 Capacitive dielectric.

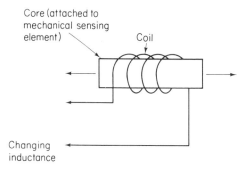

Core (attached to
mechanical sensing
element)

Coil

Changing
inductance

Figure 7-7 Inductive transduction.

toward or away from the other plate. Sometimes the moving plate is the sensing element (such as when one plate is formed by a diaphragm). In other
cases, both plates are stationary, and the change occurs in the dielectric,
such as when the dielectric is a liquid that rises and falls between the plates.
Capacitive transducers are often used in a bridge circuit similar to that of
Fig. 7-4, with an a-c excitation voltage applied.

Inductive transduction. As shown in Fig. 7-7, the measurand is converted
into a change of self-inductance of a single coil in an inductive transducer.
This change in self-inductance is usually done by displacement of the coil

core, which is linked or attached to a mechanical sensing element. Inductive transducers are often used in bridge circuits similar to that of Fig. 7-4, with an a-c excitation voltage.

Electromagnetic transduction.   As shown in Fig. 7-8, the measurand is converted into a voltage induced in a conductor by a change in magnetic flux in an electromagnetic transducer. This transducer is thus self-generating and needs no excitation voltage. Change in magnetic flux is usually accomplished by relative motion between a magnet or a piece of magnetic material and an electromagnet (which is a coil with iron or other magnetic metal core). Phonograph pickups and magnetic microphones are common examples of electromagnetic transducers. A microphone converts (transduces) sound waves into a signal voltage. A phonograph pickup converts mechanical motion (caused by the pickup moving over the record surface) into voltage.

Reluctive transduction.   As shown in Fig. 7-9, the measurand is converted into an a-c voltage change by a change in the reluctance path between two or more coils, with a-c excitation applied to the coil system, in a reluctive transducer. Generally, reluctive transducers are used in bridge circuits or as differential transformers. Change in reluctance is generally accomplished by displacement of a magnetic core (sometimes called the *armature*) linked to a mechanical sensing element. Reluctive transducers require an a-c excitation voltage.

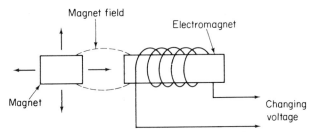

Figure 7-8   Electromagnetic transduction.

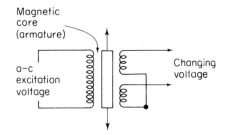

Figure 7-9   Reluctive transduction.

Piezoelectric transduction.   As shown in Fig. 7-10, the measurand is converted into a voltage generated by crystals (when mechanically stressed) in a piezoelectric transducer. In some piezoelectric transducers, the electrostatic charge between the two plates (which compress the crystals) is changed by the measurand. Mechanical stress is developed by tension, stress, compression, or bending of the crystals between the plates. In turn, these forces are exerted by the sensing element connected to the plates. Piezoelectric transducers are self-generating and thus require no excitation voltage.

Photoconductive transduction.   As shown in Fig. 7-11, the measurand is converted into a change in resistance (or the reciprocal of resistance, which is conductance) of a semiconductor material by a change in the amount of light striking the material in a photoconductive transducer. In some photoconductive transducers, the change of light is produced by a moving shutter between a light source and the semiconductor material. The shutter is linked to a sensing element. In other photoconductive transducers, the resistance change is produced entirely by a change in light surrounding the semiconductor material. Photoconductive transducers are often used in bridge circuits with d-c excitation.

Photovoltaic transduction.   As shown in Fig. 7-12, the measurand is converted into a change in the voltage generated when light strikes the junction between certain dissimilar materials in a photovoltaic transducer. Since

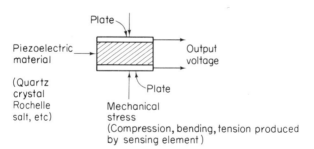

Figure 7-10   Piezoelectric transduction.

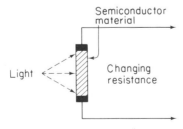

Figure 7-11   Photoconductive transduction.

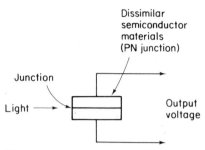

Dissimilar
semiconductor
materials
(PN junction)

Junction

Light →

Output
voltage

Figure 7-12 Photovoltaic transduction.

photovoltaic transducers are self-generating, no excitation voltage is required, so they can be used for the direct measurement of light intensity. In some transducers, the photovoltaic element is used with a fixed light source and a moving shutter operated by the measurand sensing element.

7-3 BASIC SIGNAL CONDITIONING CIRCUITS

The most common signal conditioning circuits involve some form of *resistance network;* less frequently, *inductances* or *capacitances* are used instead of resistors. In addition to these networks, there are four signal conditioning circuits or devices common to most transducers, namely, amplifiers, a-c to d-c converters, d-c to a-c converters and analog-to-digital converters. We now describe each of these networks and circuits.

7-3.1 Resistance, Inductance, and Capacitance Signal Conditioning Networks

Figure 7-13 shows the classic resistance networks found in signal conditioning circuits. These include a voltage divider, voltage drop, and bridge. The bridge also uses inductors (coils) and capacitors in place of resistances. (Likewise, the bridge principle is used in measurement systems as discussed in Chapter 8).

The *voltage divider* circuit of Fig. 7-13 consists of the transducer resistance element R, a fixed load resistance R_L, and an excitation voltage. The resistance of R is proportional to the measurand, the resistance of R_L is constant, and the output signal voltage is taken from across R_L. The output signal is proportional to the ratio of R and R_L. If R increases due to a change in the measurand, the signal voltage decreases, and vice versa. Thus, the output signal is proportional to the measurand.

The *voltage drop* circuit of Fig. 7-13 consists of the transducer resistance element R and a constant current source. The current source remains con-

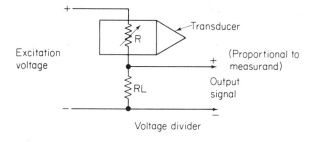

Voltage divider

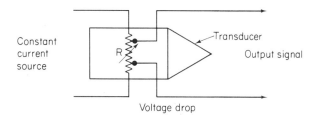

Voltage drop

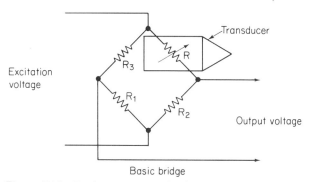

Basic bridge

Figure 7-13 Resistance networks found in signal conditioning circuits.

stant despite changes in R. Thus, since the voltage across a resistance (the output signal voltage in this case) is directly proportional to the product of current and resistance, the output signal is proportional to the value of R. If R increases due to a change in the measurand, the output voltage increases, and vice versa. Thus, the output signal is proportional to the measurand.

The *bridge* circuit of Fig. 7-13 consists of the transducer resistance element R, two fixed resistances R1 and $R2$, an adjustable resistance $R3$, and an excitation voltage. Element R forms one leg of the bridge and has a resistance proportional to the measurand. Resistor $R3$ forms the opposite leg of the bridge and is adjustable to balance the bridge. Resistors $R1$ and $R2$ complete the bridge circuit.

The output signal of the bridge circuit is proportional to the ratio of all the resistance values. In the circuit of Fig. 7-13, the output voltage equals:

$$\frac{R1}{R1 + R} - \frac{R2}{R2 + R3} \times \text{excitation voltage}$$

Thus, if all the resistances are equal, the output voltage is zero. In normal operation, $R3$ is adjusted to produce a zero output voltage when the transducer element R is at midrange or at one end of the measurand range. When the measurand changes, resistance R changes by a proportional amount, the bridge is unbalanced (the ratios of the resistors are changed), and an output signal is produced. The output signal is proportional to the measurand.

When a transducer element is capacitive or inductive, a balanced bridge can still be used. However, the excitation voltage must be alternating current. As shown in Fig. 7-14, two of the bridge legs can be capacitors (or coils) and the remaining two legs formed with fixed resistors. Or all four legs can be capacitors (or coils). No matter what form the bridge takes, the output voltage is proportional to the ratios of the elements in the legs. Usually, one or more of the elements are adjusted to produce a balance when the measurand is at midpoint or one end of its range. Any change in measurand produces a corresponding change in bridge ratios and results in a proportional signal output.

The conditioning circuits of Figs. 7-13 and 7-14 produce an output voltage that is proportional to the amplitude of the measurand or to the change in the measurand from a given level or value. The circuits can be designed to show the direction in which the measurand is changing. For example, assume that the voltage divider circuit of Fig. 7-13 is used to condition the output of a photoconductive light transducer. When the light increases, the resistance of the photoconductive element decreases and the output voltage increases. Thus, an increase in output voltage indicates an increase in light (and vice versa), and the amplitude of the output voltage is proportional to the increase in light.

This arrangement does not work well where it is necessary to sense the change in a measurand above and below a given point, that is, where it is necessary to indicate the *polarity of the measurand change*. A bridge circuit is generally used in this circumstance. Figure 7-15 shows a typical arrangement where a transducer and signal conditioning circuits are to measure linear motion by using a resistance element R with a slider arm. The moving device, mechanically coupled to the slider arm, is normally at rest in a center position. The device can be moved both right and left of the center position, and the slider arm follows.

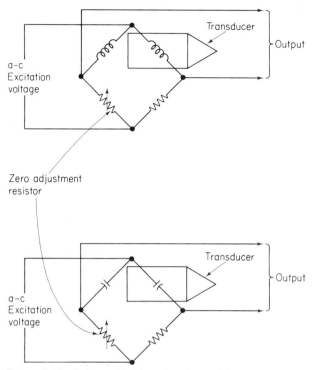

Figure 7-14 Balanced bridge signal conditioning networks for capacitive or inductive transducers.

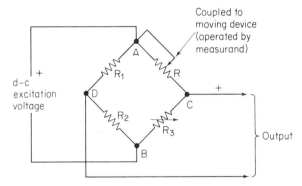

Figure 7-15 Balanced bridge signal conditioning circuit showing polarity and amount of measurand.

The resistance element R is connected as one leg of a bridge. The remaining legs are formed by resistor $R3$, which is adjustable, and fixed resistors $R1$ and $R2$. The d-c excitation voltage is applied at points A and B; the output signal is taken from points C and D. With the moving device (and the slider arm) at the center position, the resistance of R is about one-half the total resistance of R (the remaining resistance is shorted out).

Resistance $R3$ is then adjusted so that the output signal (at C and D) is zero. Assume, for example, that the excitation voltage is 10 V and that $R1$ and $R2$ are exactly equal. The voltage at point D is then 5 V. Resistor $R3$ is then adjusted so that the voltage at point C is also 5 V (the value of $R3$ equals that of R with the slider at the center position).

Now assume that the moving device and slider are moved to the right. This shorts out more of R and causes the voltage at C to rise above 5 V (there is a greater voltage drop across $R3$ and a lesser drop across R). Point C is then at a higher voltage (say 8 V), but point D remains at 5 V. Thus, there is a 3 V differential between points C and D (point C is more positive than point D, by 3 V). Thus, when point C goes positive (with respect to point D), you know that the moving device has moved to the right. The amount of differential (between C and D) indicates how far to the right the device has moved. When the moving device goes to the left of center, point C is negative with respect to D.

7-3.2 A-c to d-c Signal Conditioning

The next most common group of signal conditioning circuits are those which convert alternating current into direct current. The inductive and capacitive transducers of Fig. 7-14 require an a-c excitation signal, and the resistive transducers of Fig. 7-13 can be used with a-c excitation. When a d-c output signal is required, the signal conditioning circuits first produce an a-c output signal that is proportional to the measurand and then convert it to a d-c output signal. This conversion function is known as *rectification* or *demodulation.* (In a strict sense, rectification is the conversion of a-c into d-c, whereas demodulation implies removing an a-c component from another signal. However, the terms are used interchangeably in transducer work.)

The *solid-state diode* is the most common rectifier or demodulator used in signal conditioning circuits. Figure 7-16 shows the symbol and physical construction of a typical solid-state diode. Note that it is composed of two dissimilar solid-state materials (typically silicon or germanium crystals): P-type (positive-type) and N-type (negative-type) materials are linked together. We do not describe the structure and theory of diode materials here. For our purposes, it is sufficient to understand that current (or electrons) will flow from the N material through the junction to the P material when the P-type material (called an *anode*) is made positive with respect to the N-type material (called a *cathode*). This condition is known as *forward bias.* When the diode or rectifier is in a *reverse bias condition,* anode negative and cathode positive, no current flows.

Figure 7-17 shows how this diode principle can be used to convert a-c into d-c in signal conditioning circuits. This basic circuit, known as a *half-wave rectifier,* uses one diode and a load resistance R. When the a-c output

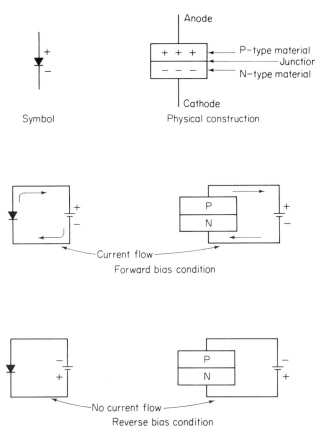

Symbol

Physical construction

Forward bias condition

Current flow

Reverse bias condition

No current flow

Figure 7-16 Solid-state diode.

signal from the transducer causes the anode to go positive, current flows through *R* in the direction indicated by the arrows. On the opposite half-cycle of the a-c input signal, the anode goes negative and there is no current flow through *R,* as shown. Thus, the current through *R* is always in one direction (it has been converted to direct current). (In a strict sense, there is a very small current flow through a diode when in the reverse bias condition. This is known as *leakage current* and can generally be ignored for practical applications.)

Figure 7-17 shows two additional circuits that use diodes for a-c to d-c conversion. The *full-wave rectifier* requires that the output signal be taken from a winding or other element with a *center-tap* connection. Because of this restriction, the *full-wave bridge rectifier* is more often used in transducer work. With either circuit, current flows through the load resistor *R,* in one direction only, on both halves of the a-c signal cycle as shown by the arrows in Fig. 7-17.

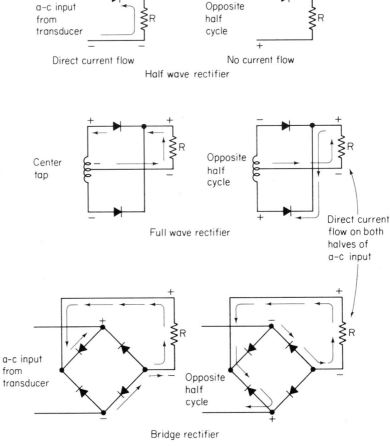

Figure 7-17 Solid-state diodes used as rectifiers to convert a-c input from transducer to d-c output.

7-3.3 *Amplifier Signal Conditioning Circuits*

Another common signal conditioning circuit is the transistor amplifier, which is used to raise the level of transducer output signals. Transducers, particularly photovoltaic and piezoelectric, produce outputs in the millivolt range, whereas a controller may require signals in the 5 V range. Transducer amplifier circuits can have *discrete* components, where each transistor and circuit part is a separate component, or *integrated circuits* (IC), where all the parts are fabricated on one "chip" of semiconductor material (usually silicon) and sealed as a single module. The trend today is to use IC amplifiers for signal conditioning of transducer outputs.

130

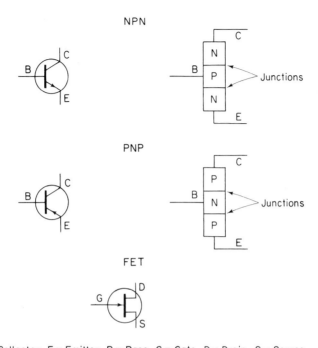

C = Collector, E = Emitter, B = Base, G = Gate, D = Drain, S = Source

Figure 7-18 Symbols for NPN, PNP, and FET transistors.

We do not go into the theory of transistors here. However, you should have a basic understanding of transistors and how they relate to signal conditioning circuits. (Note that transistors are also used extensively with the control circuits described in Chapter 9.) Figure 7-18 shows the symbols for the three common types of transistors used in control and instrumentation systems, namely, NPN, PNP, and FET (*field-effect transistor*).

The NPN and PNP are *bipolar* or *two-junction* transistors, usually made of silicon or germanium. Both types have three elements: the *collector, emitter,* and *base*. There is a junction between the collector and base and also between the emitter and base, similar to the diode junction discussed in Sec. 7-3.2. In the NPN transistor, current will flow from emitter to base when the base is made more positive than the emitter. Current will also flow from emitter to collector under this condition. If there is no voltage difference between emitter and base, or if the base is negative with respect to the emitter in an NPN transistor, no current flows. Thus, current in an emitter-collector circuit can be controlled by current in an emitter-base circuit. In effect, emitter-base current can turn a transistor on or off, or control the amount of current flow.

An emitter-base circuit is the input of a transistor, whereas an emitter-

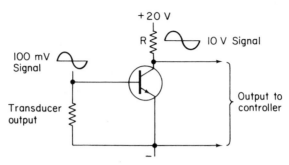

Figure 7-19 Basic signal conditioning amplifier using NPN transistor.

collector circuit is the output. An emitter-base circuit operates at low voltages compared to an emitter-collector. Typically, an emitter-base circuit operates in the 300 to 500 mV range (and can be made to operate with only a few millivolts), whereas an emitter-collector circuit can be operated with 5 V, 10, or 20 V (or even higher voltages). Thus, it is possible to control a 20 V output with an input of a few millivolts. In effect, input of a few millivolts is amplified.

As an example, assume that an NPN transistor is used as a signal conditioning amplifier in the circuit of Fig. 7-19. The transducer output appears across the emitter-base circuit. A load resistor *R* is connected across the emitter-collector circuit, and the transistor output is taken from across *R*. The resistance value of *R* is such that when maximum current (*saturation* current) flows in the emitter-collector circuit, the voltage drop across *R* is one-half the total power supply voltage applied to the emitter-collector circuit. That is, when the transistor is "full on" the collector voltage drops from 20 V to 10 V (when the transistor is "off," no emitter-collector current flows and the collector is at the power supply voltage of 20 V). Thus, there is a 10 V swing at the transistor output produced by a few millivolts change at the input—assuming that a 100 mV input signal (from the transducer) produced a 10 V output signal (from the transistor to the controller). This is a voltage gain of 100.

When considerable gain is required in a signal conditioning circuit, several transistors can be operated in *cascade,* where the output of one transistor feeds the input of another, etc. Typically, no more than three transistors are used in cascade for voltage amplification. Note that a PNP transistor operates the same as an NPN, except that power supply polarities are reversed. In a PNP, emitter-collector current flows when the base is made negative with respect to the emitter.

Internal operation of the *field-effect transistor* (or FET) shown in Fig. 7-18 is quite different from that of either an NPN or PNP. However, the overall effect is basically the same insofar as signal conditioning amplifiers

132

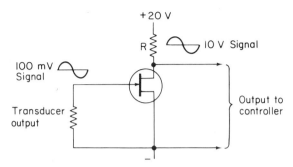

Figure 7-20 Basic signal conditioning amplifier using a FET (field effect transistor).

are concerned. As shown in Fig. 7-20, an FET can be connected in a circuit similar to that of an NPN or PNP. However, the input signal from the transducer is connected to the *gate* circuit. Output from the FET is taken from a *source* and *drain* circuit. Again, a small voltage change at the input (gate) produces a large voltage change at the output (source, drain), resulting in amplification of the signal.

For a more comprehensive discussion of transistors, your attention is called to the author's *Handbook for Transistors* (Prentice-Hall, Inc., Englewood Cliffs, N.J. 07632, 1976).

7-3.4 D-c to a-c Signal Conditioning

Sometimes, it is necessary to convert direct current into alternating current. For example, assume that all but one of the transducers in a control system require d-c excitation voltages. The remaining one transducer requires an a-c excitation, and thus would require a separate power supply (which might be uneconomical or otherwise impractical). This problem can be solved by using a signal conditioning circuit that converts a d-c power supply (used for all transducers) into an a-c excitation voltage for the one "odd-ball" transducer.

Figure 7-21 shows the circuit of a reluctive transducer that must be operated from a d-c supply and must produce a d-c output signal. The transducer element is a form of linear variable differential transformer, discussed in Sec. 2-2, and produces an a-c output proportional to linear motion. This a-c signal output is converted by the rectifier or demodulator circuit, as discussed in Sec. 7-3.2. The a-c excitation voltage for the transducer element is provided by the d-c to a-c *converter* or *inverter,* which consists essentially of two PNP transistors, a transformer, and two resistors.

Note that both transistors *A* and *B* are placed in a position to be turned on when the power supply is applied. That is, the emitters are connected to the positive end of the +5 V supply, whereas the bases are connected to the

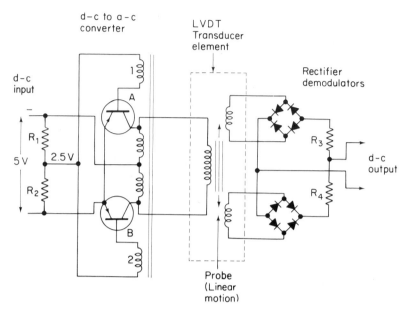

Figure 7-21 Reluctive transducer with d-c input and output.

junction of $R1$ and $R2$ through the transformer windings. This junction is at about $+2.5$ V (assuming that $R1$ and $R2$ are equal), making the bases negative with respect to the emitters. This forward bias turns on the PNP transistors, and emitter-collector current flows through the center-tapped winding of the transformer, producing lines of force. These lines of force from the center-tapped winding are picked up by the base windings 1 and 2, which are out-of-phase, so that one base receives a positive voltage when the other base is being driven negative, and vice versa.

Since transistors and transformer windings are never perfectly balanced, one transistor is driven into saturation, while the opposite transistor is cut off. When a transistor reaches saturation and no further increases in current can occur, the lines of force collapse. This collapse produces a reverse in the voltages produced by the base windings, and the transistors switch states. The repeated rise and fall of the lines of force around the center-tapped winding generate an a-c voltage in the transformer secondary. This a-c voltage is used as the excitation voltage for the transducer element.

Note that the rectifier circuit shown in Fig. 7-21 is somewhat different from that discussed in Sec. 7-3.2. There are two full-wave bridge rectifiers in the Fig. 7-21 circuit, one rectifier for each of the transducer element secondary windings. Each rectifier produces a d-c voltage proportional to the a-c voltage on the corresponding winding. These d-c output voltages appear across resistors $R3$ and $R4$. If the moving device (being sensed by the transducer) is at the center position, the outputs from both secondaries (and both rectifiers) are equal, and the voltage drops across $R3$ and $R4$ are equal but of opposite polarity (since the rectifier diodes are arranged to produce opposite polarities). Under these "at rest" conditions, the signal output is zero.

When the moving device moves in one direction from center, the voltage drops across $R3$ and $R4$ are unequal, and the difference appears as a d-c voltage. When the device is moved in the opposite direction from center, the difference appears as a d-c voltage of the opposite polarity. The polarity of the output signal indicates the direction of motion, whereas the output signal amplitude indicates the amount of motion away from center.

7-3.5 Analog-to-Digital Signal Conditioning

Transducers produce a voltage which is an *analog* of the condition or force they transduce. That is, the voltage is analogous or representative of the force or condition. Many controllers will accept these analog signals in voltage (or current or resistance) form. This is particularly true of simple controllers and most older controllers. Modern controllers often operate by digital principles. In effect, they are digital computers of a sort and require digital inputs. Some digital controllers contain analog-to-digital (A/D) conversion circuits and will accept analog voltage inputs. Other controllers use only digital inputs, thus requiring the transducer output to be converted or conditioned.

We now describe the basic principles of A/D conversion. You must understand the basics of digital electronics (including pulses and the binary number system) to understand operation of an A/D converter. The author's *Logic Designer's Manual* (Reston Publishing Company, Reston, VA., 22090, 1977) provides a full description of these subjects. If you do not want to get into digital electronics, you can simply accept that an A/D converter is a device (possibly an integrated circuit packaged within a transducer) that converts a measured voltage into a series of digital pulses, representing the voltage in binary form.

Analog-to-digital conversion. One of the most common methods of direct A/D conversion between tranducer output and controller input involves the use of a converter that operates on a sequence of *half-split, trial-and-error steps.* This method produces output voltage pulses, arranged in binary code, that are representative of the input voltage from a transducer.

The heart of such a converter is the *conversion ladder* such as shown in Fig. 7-22. The ladder provides a means of implementing a four-bit binary coding system and produces an output that is equivalent to the switch positions. The switches can be moved to either a 1 or a 0 position, which corresponds to a four-place binary number. The output voltage to the controller describes a percentage of the full-scale reference voltage, depending

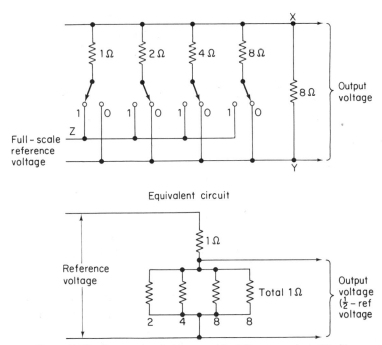

Figure 7-22 Conversion ladder used in A/D conversion circuit.

on the binary switch positions. For example, if all switches are in the 0 position, there is no output voltage. This produces a binary 0000, represented by 0 V.

If switch A is at 1, and the remaining switches are at 0, this produces a binary 1000 (decimal 8). Since the total in a four-bit system is 16, 8 represents one-half of full scale. Thus, the output voltage is one-half of the full-scale reference voltage, which is done as follows:

The 2-, 4- and 8-ohm switch resistors and the 8-ohm output resistor are connected in parallel. This produces a value of 1-ohm across points X and Y. The reference voltage is applied across the 1-ohm switch resistor (across points Z and X) and the 1-ohm combination of resistors (across points X and Y). In effect, this is the same as two 1-ohm resistors in series. Since the full-scale reference voltage is applied across both resistors in series, and the output is measured across only one of the resistors, the output voltage is one-half the reference voltage. Note that the reference voltage is generally equal to the full-scale output of the transducer.

In a practical A/D converter, the same basic ladder is used to supply a comparison voltage to a *comparison circuit,* which compares the transducer output voltage to be converted with the binary-coded voltage from the ladder. The resultant output of the comparison is a series of pulses, in binary code form, representing the voltage to be converted.

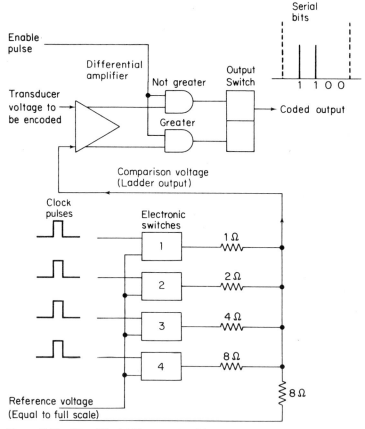

Figure 7-23 Simplified A/D converter (binary encoder) using the four bit system.

In a practical circuit, the mechanical switches shown in Fig. 7-22 are replaced by electronic switches. When a switch is "on," the corresponding ladder resistor is connected to the reference voltage. When "off," the resistor is disconnected from the reference voltage. The switches are triggered by four pulses (representing a four-bit binary number) from a system clock. An enable pulse is used to turn the comparison circuit on or off, so that as each switch is operated a comparison can be made of the four bits.

Simplified A/D operating sequence. Figure 7-23 is a simplified block diagram of an A/D converter. Here, the reference voltage (equal to the full-scale output of the transducer) is applied to the ladder through the electronic switches. The ladder output (comparison voltage) is controlled by switch positions which, in turn, are controlled by the pulses.

We now outline the sequence of events necessary to produce a series of four pulses (arranged in binary form) that describe the voltage from the transducer as a percentage of full scale (in one-sixteenth increments). Assume that the transducer voltage is three-fourths of full scale (or 75%). Also assume that the full-scale output from the transducer is 10 mV and that the transducer shows an actual output of 7.5 mV.

Note that the ladder output is applied to a comparison circuit consisting of a differential amplifier, two gates, and an output switch. The balance of the amplifier is set so that its output is sufficient to turn on one gate and turn off the other gate, if the ladder voltage is greater than the transducer voltage. Likewise, the differential amplifier will reverse the gates if the ladder voltage is not greater than the transducer voltage. When the "not greater" gate is on, the output switch produces a pulse (which is equal to a binary 1). When the "greater" gate is on, the output switch does not produce a pulse (which is equal to a binary 0). Both gates are turned on by pulses from the system clock.

When pulse 1 arrives, ladder switch 1 is turned on and the remaining ladder switches are off. The ladder output is a 50% voltage that is applied to the comparator circuit. In our example (with the transducer at 75% of full scale), the ladder output is less than the transducer voltage when pulse 1 is applied to the ladder. As a result, the "not greater" gate turns on, and the output switch produces a pulse (binary 1). Thus, for the first of the four pulses in the binary number the output is 1.

When pulse 2 arrives, switch 2 is turned on and switch 1 remains on. Both switches 3 and 4 remain off. The ladder output is now 75% of full-scale voltage and equals the transducer voltage. However, the ladder output is still not greater than the transducer voltage. Consequently, when the gates are enabled, they remain in the same position as before ("not-greater" gate on, "greater" gate off). The output is again a 1 (the second of the four pulses is a 1).

When pulse 3 arrives, switch 3 is turned on. Switches 1 and 2 remain on, while switch 4 is off. The ladder output is now 87.5% of full-scale voltage and is thus greater than the transducer voltage. As a result, when the gates are enabled, they reverse. The "not-greater" gate turns off, and the "greater" gate turns on. The output is then a 0 (no pulse).

When pulse 4 arrives, switch 4 is turned on. All switches are now on. The ladder is now maximum (full scale) and thus is greater than the transducer voltage. As a result, when the gates are enabled, they remain in the same condition ("not-greater" off, "greater" on), and the output is a 0 (no pulse).

The four pulses from the A/D circuit are 1, 1, 0 and 0, or 1100. This is a binary 12, which is 75% of 16. Thus, the transducer voltage (7.5 mV, or 75% of 10 mV) is converted to pulses 1100, which represent 75% of full scale to the controller.

In a practical circuit, when the fourth pulse has passed, all the switches are reset to the off position, thus placing them in a position to spell out the next four-pulse binary word.

7-4 THICKNESS SENSORS

There are a number of sensors used to measure the thickness of materials. We describe now some typical examples.

7-4.1 Measuring Thickness of Magnetic Materials

Figure 7-24 shows a sensor system used to measure the thickness of magnetic material. The sensor element consists of a coil of wire wound on one arm of a U-shaped core of magnetic material. The open end of the core is placed upon the magnetic material under test, thus completing a magnetic circuit. Since the material being measured is part of the magnetic circuit, the inductance of the coil is determined (in part) by the thickness of the material. Thus, by measuring the inductance of the coil, the thickness of the material is determined.

The coil inductance can be measured on a bridge, as discussed in Chapter 8. As an alternate, the coil can be made part of an oscillator, the frequency of which depends on the coil inductance, as shown in Fig. 7-24. By measuring oscillator frequency, the coil inductance (and thus the thickness of the material is determined. Frequency measurement is discussed in Chapter 8.

7-4.2 Measuring Thickness of Insulated Materials

Figure 7-25 shows a sensor system used to measure the thickness of insulator material such as nonconducting plastic. With this system, metal plates are placed at either side of the material. Thus, the material and plates form a capacitor, with the material acting as a dielectric. Since the capacitance depends (in part) on the thickness of the dielectric, material thickness is determined by measuring the capacitance.

The capacitance can be measured on a bridge, as discussed in Chapter 8. As an alternate, the capacitance can be made part of an oscillator, the frequency of which depends on the capacitance as shown in Fig. 7-25. By measuring oscillator frequency, the capacitance (and thus the thickness of the material) is determined. Frequency measurement is discussed in Chapter 8.

7-4.3 Measuring Thickness with Ultrasonic Vibrations

Another type of thickness sensor uses ultrasonic vibrations. These are mechanical vibrations that use a gas, liquid, or solid as a medium and whose frequencies are beyond the audio range (that is, more than about 15,000

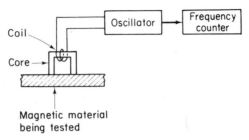

(Coil forms part of oscillator frequency determining circuit)

Figure 7-24 Sensor system used to measure thickness of magnetic material.

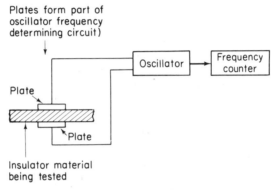

Plates form part of oscillator frequency determining circuit)

Figure 7-25 Sensor system used to measure thickness of insulator material.

vibrations per second). The vibrations are produced by a transducer which converts the electrical output of an oscillator to ultrasonic vibrations of corresponding frequencies.

There are two kinds of ultrasonic transducers. One is the *magnetostrictive* type, which consists of a metal rod placed in a coil that is driven by oscillator signals as shown in Fig. 7-26(a). As a result of the alternating magnetic field generated by the coil, the rod alternately becomes a little longer and little shorter at the frequency of the oscillator. Since one end of the rod is fixed, the opposite end pushes and pulls on a plate or *diaphgram,* producing ultrasonic sound waves. In effect, the magnetostrictive transducer operates somewhat like a loudspeaker but does not use exactly the same principle of operation. Sound is discussed further in Sec. 7-8.

The other, more common type of ultrasonic transducer consists essentially of a *piezoelectric crystal,* such as quartz, shown in Fig. 7-26(b). As discussed in other chapters, piezoelectric materials produce a voltage when

140

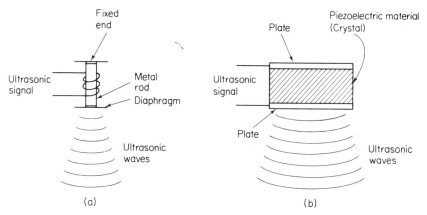

Figure 7-26 Ultrasonic transducers: (a) magnetostrictive; (b) piezoelectric.

they are compressed or mechanically vibrated. The reverse is also true. When a voltage is applied to a piezoelectric material, it will compress or expand. If the voltage is alternating at an ultrasonic frequency, the piezoelectric material will compress and expand at the same ultrasonic frequency. In effect, the material will vibrate, and these vibrations can be transferred to a diaphragm to produce ultrasonic sound waves.

Figure 7-27 shows one way to use ultrasonic vibrations for the measurement of material thickness. (Note that this same technique can be used to measure the contents or the level of liquid in a tank.) For thickness measurement, the transducer is placed on top of the material, so that the ultrasonic vibrations pass through the material to a flat surface. When vibrations strike the other side of the material, they are reflected back to the transducer from the surface. The time it takes for the vibrations to make one round trip depends, among other factors, on the thickness of the material.

If the thickness of the material is such that the time required for one round trip is equal to the time of one cycle of the ultrasonic vibration, a condition of *resonance* is produced. Resonance occurs when the vibrations return at the exact time to reinforce outgoing vibrations. The exact timing depends on thickness and oscillator frequency. At resonance, there is a sudden change in the load that the transducers offer the oscillator. The change in load produces a corresponding change in the oscillator current (the collector current of a transistor oscillator). The change can be indicated by a meter placed in the oscillator circuit.

By noting the frequency of the oscillator at the point where the current change takes place, the time required for a round trip of the ultrasonic vibrations (and thus the thickness of the material) is determined. The frequency of the oscillator may be varied by means of a capacitor (or possibly

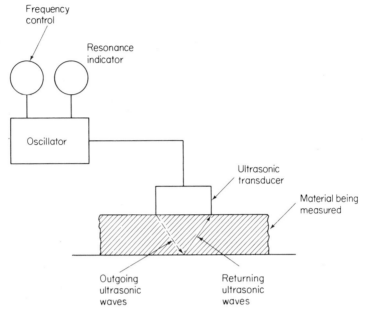

Frequency control

Resonance indicator

Oscillator

Ultrasonic transducer

Material being measured

Outgoing ultrasonic waves

Returning ultrasonic waves

Figure 7-27 Using ultrasonic soundwaves (vibrations) to measure thickness of materials.

an inductor) operated by a tuning control. When the condition of resonance is reached (by noting the change in oscillator current), the reading of the tuning control (which may be calibrated in units of thickness) will indicate the thickness of the material.

7-4.4 Measuring Thickness with Radiation Sensors

The radiation sensors described in Chapter 5 can be used to measure thickness. Radiation technique is particularly useful when manufacturing sheets of material made of metal, plastic, paper, rubber, and artificial leather, where it is important to keep check on the thickness of sheets as they are wound rapidly into rolls.

Figure 7-28 shows one system for measuring thickness by using a radiation sensor. Here, the sheet material passes between a lower chamber containing a source of radiation (X rays, gamma rays, beta rays) and an upper chamber containing one of the radiation sensors described in Chapter 5. A continuous stream of radiation is emitted from a window in the lower chamber, passes through the material whose thickness is being measured, and strikes the sensor in the upper chamber. As the radiation passes through the sheet material, some of the radiation is absorbed and, as a result, the radiation reaching the sensor is less intense. The amount of absorption

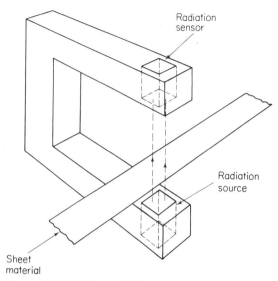

Figure 7-28 Using radiation sensor to measure thickness of materials.

depends on the material's density and thickness of the sheets. If the material and density are kept constant, the amount of radiation absorption varies directly with the thickness of the sheet.

Knowing the original intensity of the radiation, and the intensity of the radiation reaching the sensing device, the amount of absorption is determined. In this way, the thickness of the moving sheets becomes known. Of course, the same method may be used to measure the thickness of stationary material as well. In all these systems, the output of the sensor can be read on a meter or other type of indicator.

Gamma rays, because of their greater penetrating ability are used to measure thicker metallic materials. Beta rays, because of their lesser penetrating ability, are generally used for thinner nonmetallic materials. X rays can be used for either thickness. The frequency of X rays depends on the anode voltage of an X-ray tube. The higher the anode voltage, the higher the frequency of the X rays produced. X rays having relatively low frequencies are known as *soft X rays.* Those with higher frequencies are known as *hard X rays.* Hard X rays are much more penetrating than soft X rays, and the higher the frequency, the more penetrating the X-ray. Thus, when measuring thick materials, the more penetrating hard X rays or gamma rays are used. For thin or nonmetallic materials, the less penetratiang soft X rays or beta rays may be used.

The same principles involved in radiation-type thickness sensors can be used in the examination of metal castings for hidden flaws. Radiation absorption by such flaws generally is less than their absorption by the surrounding material. Thus, if a sensor is passed over the surface of a casting, the presence of a flaw will be revealed by a sudden drop in the amount of absorption.

7-5 PROXIMITY SENSORS

The proximity sensor is a device used to detect the proximity (or the absence) of a body. The speed-of-rotation sensor described in Sec. 2-4 is a form of proximity sensor. This sensor consists of a coil wound on a small permanent magnet. When a metal object passes near the sensor (called a *magnetic pickup*), the magnetic field is changed, thus inducing a voltage pulse in the coil.

A speed-of-rotation sensor may be used for many applications, such as counting metallic objects moving past the sensor on a conveyor belt. This sensor can also be constructed in the form of a tube and used to count small metallic objects as they fall, one at a time, through the tube as shown in Fig. 7-29. Each falling part produces an electrical pulse that can be counted on an electronic counter such as described in Chapter 8.

Also as shown in Fig. 7-29, a hollow magnetic pickup can be placed around a drill used for an automatic machining process. When the drill rotates, a voltage is induced in the coil. Should the drill break, absence of rotation within the pickup is sensed by the lack of induced voltage, and the machine shuts off automatically.

7-5.1 Oscillator-Type Proximity Sensors

An r-f (radio-frequency) oscillator can be used as a sensing device to detect the presence (or absence) of metallic objects. An example is shown in Fig. 7-30, where the coil of an oscillator circuit is placed in a special housing and connected to the oscillator parts by means of a cable. The coil then becomes a *pickup* to sense the presence of metals. Normally, the oscillator is in operation, and the flow of base bias current causes the pickup coil to be surrounded by a magnetic field. As a metal object enters the field around the coil, the object absorbs enough energy from the oscillator to stop oscillation. This changes the collector current of the transistor sufficiently to operate a relay or other control device. Relays are discussed in Chapter 9.

Figure 7-31 shows a variation of the oscillator-type proximity sensor used for aligning elevators at floor levels. An oscillator is mounted at each floor and an iron vane is attached to the bottom of the elevator cage. When the cage is at the proper floor level, the metal vane is positioned between coils $L1$ and $L2$ of the oscillator, thus shielding the coils from each other and eliminating the feedback between coils and stopping oscillations. The resulting change in oscillator current then operates a control device (such as a relay) that stops the elevator motor, making it possible to open the elevator door.

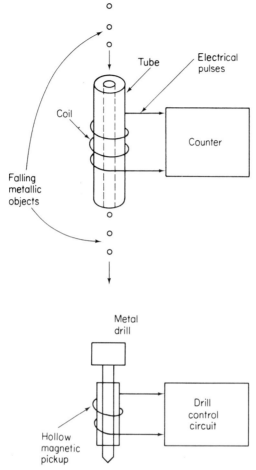

Figure 7-29 Proximity sensors used to count falling metal objects and detect absence of metal drill.

Figure 7-32 shows another example of an oscillator-type proximity sensor. Oscillations are maintained as a result of feedback that compensates for losses in the coil-capacitor circuit. The circuit is adjusted so that there is just enough feedback to maintain oscillation. Should any other load be placed on the circuit, oscillations stop.

If a person approaches the metal plate connected to the oscillator circuit, the person's body forms one plate of a "capacitor" consisting of the metal plate, the body, and the air between (which acts as a dielectric). Thus, a coupling "capacitor" is formed that couples the oscillator circuit to the body, and so to ground. As a result, an additional load is placed on the oscillator and, if the load is large enough, oscillations stop.

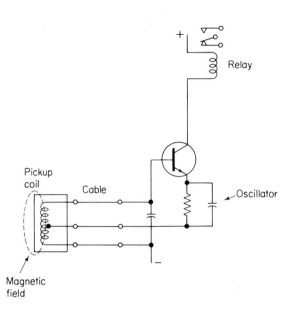

Figure 7-30 Basic oscillator-type proximity sensor.

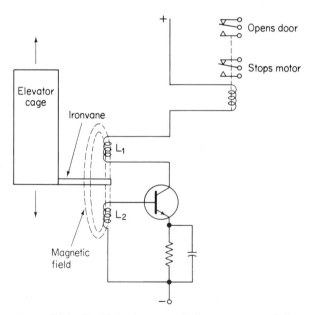

Figure 7-31 Oscillator-type proximity sensor used for aligning elevators at floor levels.

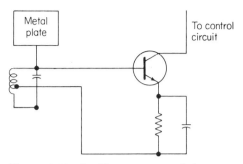

Figure 7-32 Oscillator-type proximity sensor using metal plate to stop oscillation.

This kind of oscillator proximity sensor is sometimes called a *capacitive relay* and is frequently used for advertising displays. A metal plate is glued to the inside of a glass store window, and a passer-by is requested to place his or her hand near the opposite side of the window. The glass acts as a dielectric to complete the "capacitor," the oscillator is grounded through the passer-by's body, and oscillations stop. As a result there is a change of oscillator current, thus causing a relay to operate some device, such as a lamp or motor, to be turned on or off. A capacitive relay can also be used as a door opener or burglar alarm or as a safety device, should the operators of machines place their hands into some dangerous area, or to detect and count metal objects passing on a conveyer belt.

7-6 DENSITY AND SPECIFIC GRAVITY SENSORS

The density of a substance is the mass or weight of a substance per unit volume. For example, the density of water is approximately 62.5 lb/cu ft. Density is frequently expressed in terms of *specific gravity,* which is the ratio between the weight of a substance and the weight of an equal volume of some other substance (usually water) that is used as a standard. As an example, lead weighs 687 lb/cu ft. Thus, the specific gravity of lead is a little over 10 when compared with water.

Many industrial processes depend, in part, on the densities of liquids. An obvious method for measuring density is to weigh a known volume of a liquid. A simpler method is to use the *hydrometer,* which is a calibrated float. The greater the density of a liquid, the less the float will sink into the liquid. Thus, the density can be found by noting the calibration on the float corresponding to the liquid surface. This is the method used to find the specific gravity of acid in a battery.

A variation of this hydrometer principle can be used when the density of a liquid must be monitored continuously. Such a density sensor system is shown in Fig. 7-33, where a hollow glass float, weighted at the bottom so that it will float upright, is immersed in the liquid to be monitored. The denser the liquid, the more the float will be buoyed up and the higher the float will ride. The less dense the liquid, the lower the float will sink. The float is contained in a chamber that is partially filled with a liquid, as the liquid passes through. Farther up in the chamber, above the liquid level, a light from a lamp passes through a slit in the chamber side to fall upon a photocell (Chapter 5) mounted on the opposite side of the chamber. The top portion of the float stem is opaque (so that no light will pass through). As the float rises in the liquid, the float stem passes in front of the slit, thus cutting off the light to the photocell. The position of the stem determines the amount of light that can pass and so governs the output of the photocell. The photocell output is thus inversely proportional to the liquid density.

Figure 7-34 shows another variation of the hydrometer principle used for density sensor. Here, a float is weighted by a chain to sink to a predetermined depth in the liquid, under test, flowing through a glass chamber. A rod attached to the top of the float carries the movable core of a linear variable differential transformer (Chapter 2). The primary and secondary windings of the transformer are wound around the outside of the chamber. (Because the chamber is made of glass, it does not interfere with the transformer action.)

Normally, the float is adjusted so that the transformer core is at the null position and the output voltage is zero. Should the density of the liquid increase, the float moves higher, the core is raised, and a differential output

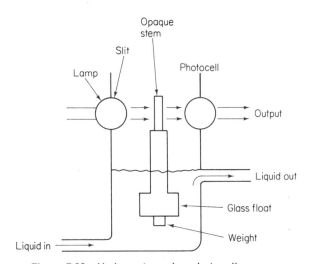

Figure 7-33   Hydrometer using photocell sensor.

voltage is generated. The amplitude of the differential voltage is proportional to the rise of the core, and the phase is dependent on the direction of the core motion. Should the liquid density decrease, the float sinks deeper, the core is lowered, and a differential voltage of opposite phase (and proportional amplitude) is generated. In either case, the output voltage phase indicates whether the liquid is more or less dense than normal, whereas the voltage amplitude indicates the amount of density change or deviation from normal. (The more the density differs from normal, the greater the voltage amplitude.)

A variation of the thickness sensors described in Sec. 7-4 can be used to measure density. With a thickness sensor, it is assumed that the density of the material under test is uniform throughout. The output signal is thus determined by thickness. One the other hand, if the material thickness is kept constant, any variations in output voltage are produced by variations in density.

For example, assume that a radiation sensor (Chapter 5) is placed on one side of a material (of a given thickness or size), with a radiation source at the opposite side. The denser the material, the more the material will absorb the radiation and the smaller will be the reading of the radiation sensor. By comparing these readings with the readings obtained from material of standard density and thickness, the density of the material under test may be determined. If liquids are involved, they may be contained in a chamber of standard diameter, and their absorption of radiation compared to that of a standard liquid in a similar chamber.

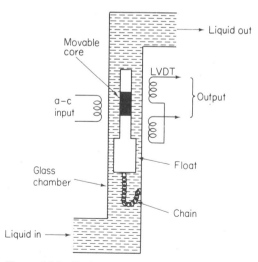

Figure 7-34 Hydrometer using LVDT to indicate density.

7-7 CHEMICAL SENSORS

There are many control and instrumentation systems devoted to the measurement of chemical content. For example, there may be a need to sense the purity or acidity of a solution or to detect pollutants in the atmosphere and give warning when they reach dangerous proportions. Typical examples of chemical sensors include the *pH (or acidity/alkalinity) sensor,* and the *thermal-conductivity gas analyzer,* which we now describe.

7-7.1 pH Sensor (Acidity Alkalinity Sensor)

Many industrial processes require that the acidity or alkalinity of solutions be measured and controlled if the processes are to proceed effectively and efficiently. The degree of acidity or alkalinity of any aqueous solution is determined by the relative concentrations of *hydrogen* and *hydroxyl ions* in that solution.

When hydrogen ions predominate, the solution is acidic. If hydroxyl ions are in the majority, the solution is alkaline. Since the produce of hydrogen-ion and hydroxyl-ion concentrations in any such solution is a constant value, measurement of hydrogen-ion concentration indicates not only the acidity of a solution, but its effective alkalinity as well.

Hydrogen-ion concentration is measured on a scale (the *pH scale*) that ranges from 0 to 14. On this scale, neutrality is 7, the value obtained when the hydrogen and hydroxyl ions are equal and balance each other, as is the case with pure water. As the solution becomes more alkaline, the pH scale reading increases above 7. As the solution becomes more acid, the pH scale reading decreases below 7.

A pH measurement is obtained by immersing a pair of electrodes in the solution to be measured and measuring the voltage across the electrodes as shown in Fig. 7-35. The action is somewhat similar to that of a voltaic cell, where a pair of dissimilar electrodes are immersed in an electrolyte.

In the pH cell, the *reference electrode* is at a constant voltage or potential, regardless of the pH of the solution under test. The potential of the *measuring electrode* is determined by the pH of the solution. Thus, the potential difference between the two electrodes depends on the pH of the solution.

The reference electrode is made of glass and consists of an inner assembly containing a solution of calomel (mercury chloride) and mercury. This assembly is surrounded by a larger glass tube, and the space between the two contains a saturated solution of potassium chloride. A tiny opening in the bottom of the electrode permits the potassium chloride to diffuse very slowly into the solution under test. In this way, electrical contact is made between the solution under test and the calomel solution of the electrode.

The measuring electrode (sometimes called the *glass electrode*) is

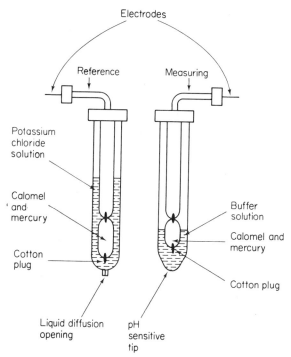

Figure 7-35 pH (acidity/alkalinity) sensor.

somewhat similar to the reference electrode. However, the mercury calomel element is surrounded by a buffer solution of constant pH as shown in Fig. 7-35. The bottom of the outer tube has no opening. Instead, the outer tube tapers down to a tip made of thin glass of special composition. At this tip, a potential difference is developed between the buffer solution and the solution under test, because of the differences in the pH of the two solutions. Since the pH of the buffer solution is constant, the net potential of the measuring electrode is a function of the pH of the solution being tested.

Both electrodes are mounted to form an assembly that is inserted into the solution being tested. The difference of potential between the electrodes, which is a function of the pH of the solution under test, can be measured by a voltmeter (Chapter 8) calibrated in units of pH. As an alternate, the difference in potential can serve as a control signal.

7-7.2 Thermal-Conductivity Gas Analyzer

The thermal-conducivity gas analyzer is one of the instruments used when the chemical purity of a gas is to be measured. One common use for such analyzers is to measure air pollution or smog. Operation of this type of analyzer is based on the difference in heat conductivity among gases. Figure 7-36 shows the basic circuit for a thermal-conductivity gas analyzer.

Note that a balanced bridge circuit is formed, with two sensing resistors acting as balanced legs of the bridge. One sensing resistor is surrounded by the gas to be analyzed, whereas the other resistor is surrounded by a reference gas (such as oxygen, pure atmosphere, etc.). Where practical, both gases are maintained at the same pressure, water content, etc.

The bridge is first balanced by exposing both resistors to the same gas. Current flows through both sensing resistors, as well as through resistors $R1$ and $R2$. Resistor $R2$ is adjusted for a "balance" or "zero set" condition on the meter. Then the resistors are exposed to the reference sample and gas sample. If the gas sample contains elements having a different thermal conductivity than the reference sample, the bridge will be unbalanced. In some cases, the meter reads out in terms of thermal conductivity where, as in other analyzers, the indication is on a GO—NO GO or GOOD—BAD basis.

7-8 SOUND SENSORS

Since sound waves are essentially mechanical vibrations, many of the vibration transduction elements discussed in Chapter 2 can also be used to measure sound. Of course, there must be some form of sensing element to

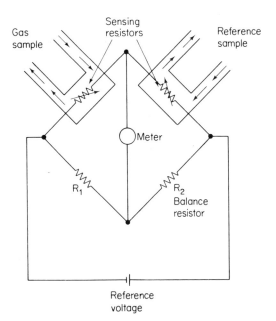

Figure 7-36 Basic thermal-conductivity gas analyzer.

detect the mechanical sound waves. The most common sound sensor is the *microphone*. However, the physical form of sound sensors used in control systems often are quite different from voice microphones, even though the transduction methods are essentially the same.

For most laboratory or control and instrumentation applications, sound is measured in units of *sound pressure,* which is the total instantaneous pressure at a given point, in the presence of a sound wave, minus the static pressure at that point. Sound-pressure transducers differ from microphones. A sound-pressure transducer provides an output in response to sound pressure which is to be measured. The microphone provides an output in response to sound waves which are not necessarily associated with a measurand.

Sound pressure is expressed in terms of *sound-pressure level,* or SPL (sometimes L_p). SPL is 10 times the *logarithmic ratio* of the *mean-square sound pressure* to a *mean-square reference pressure*. SPL is usually expressed in *decibels* as 20 times the logarithm (base 10) of the ratio of the root-mean-square (RMS) sound pressure to an RMS reference pressure, or

$$SPL = 20 \log_{10} \frac{p \text{ (RMS)}}{p_{ref} \text{ (RMS)}}$$

The reference pressure is usually considered as 2×10^{-4} dynes/cm², or 2×10^{-4} microbar. A *dyne* is the amount of force that causes a mass of 1 g to alter its speed by 1 cm/sec for each second during which the force acts. A *microbar* is one-millionth of a bar. One bar is the normal or standard atmospheric pressure at sea level.

The decibel (dB) has been widely adopted in electronics to express logarithmically the ratio between two powers, voltage levels, or sound power levels. A decibel is one-tenth of a *bel* (the bel is too large for most practical applications). The use of the decibel, which is a logarithmic unit, permits a closer approach to the reaction of the human ear. This is because the response of the human ear to sound waves is approximately proportional to the logarithm of the energy of a sound wave and is not proportional to the energy itself. The impression gained by the human ear as to the magnitude of sound is roughly proportional to the logarithm of the actual energy contained in the sound. For example, the change in gain of an electronic amplifier expressed in decibels provides a much better index of the effect of a sound upon the ear than if expressed as a power or voltage ratio.

7-8.1 Sound-Sensing Elements

The sensing element used in most microphones and sound-pressure transducers is the *flat diaphragm*. Typically, this diaphragm is a circular flat plate supported continuously around its edge. There is a variety of diaphragms used in microphones and sound-pressure transducers.

However, all sound-pressure transducers have certain construction elements in common. For example, sound-pressure sensing elements are usually constructed in a *gage-pressure* configuration (Gage pressure is discussed in Sec. 3-2.2) In a typical sound-pressure transducer the ambient pressure is admitted to the reference side (inside of case) so that sound pressure is measured with respect to ambient static pressure, while static pressures acting on the outer and inner diaphragm surfaces are equalized. This is shown in Fig. 7-37. The gate vent (case opening) also acts as a low-pass filter acoustic leak, which prevents access of the sound to the reference side of the diaphragm. Some pressure transducer sensing elements use a sealed-reference differential-pressure configuration where the reference side of the diaphragm is sealed and (possibly) partly evacuated.

7-8.2   Sound Transducing Methods

There are five commonly used sound transduction methods: capacitive, inductive, electromagnetic, piezoelectric, and reluctive. These same methods are used for both microphones and sound-pressure transducers. We now describe how these methods are used for sound measurement systems.

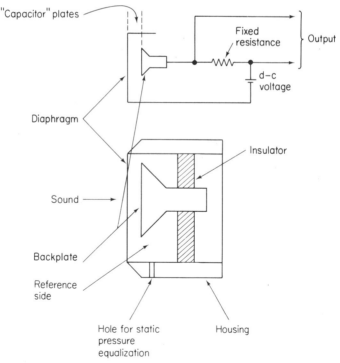

Figure 7-37   Basic capacitive sound-pressure transducer.

Capacitive-type sound transducers. Figure 7-37 shows the construction of a capacitive sound-pressure transducer. This same basic construction is also found in the once popular "condenser microphone." The diaphragm acts as one plate of a capacitor, with the backplate forming the other capacitor plate. A fixed d-c voltage is applied across the two plates through a fixed resistance. Thus, a constant charge is maintained on the plates. Changes in capacity due to diaphragm deflection (caused by sound waves striking the diaphragm) produce a change in voltage across the plates. The changing voltage produces an output that is proportional to the sound waves. (The changes occur at the sound wave frequency, and the output is directly proportional to the sound wave amplitude.) Generally, the output of a capacitive-type sound transducer (and a condenser microphone) is amplified before it is applied to a system.

Inductive-type sound transducers. Figure 7-38 shows the construction of an inductive sound-pressure transducer. This technique is not generally used for microphones. The inductive-type sound transducer contains a diaphragm made of some ferromagnetic material (material containing iron and capable of being permanently magnetized). The diaphragm changes the self-inductance of a coil when the gap between diaphragm and coil is changed by sound waves. As shown, this type of sound-pressure transducer is often used in a bridge circuit, requiring an a-c excitation voltage.

Electromagnetic-type sound transducers. Figure 7-39 shows the construction of an electromagnetic-type sound transducer. This is a form of "dynamic microphone" and is a self-generating device requiring no excitation voltage. The diaphragm is connected to a moving coil wound concentrically over a permanent magnet. Sound waves striking the diaphragm move the coil through a magnetic field produced by the permanent magnet.

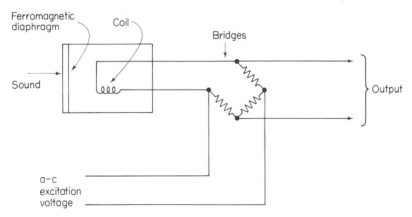

Figure 7-38 Basic inductive-type sound transducer.

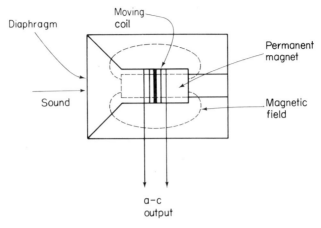

Figure 7-39 Basic electromagnetic-type sound transducer.

This generates an a-c signal in the coil that is proportional to the frequency and amplitude of the sound waves. The electromagnetic principle is not commonly used for sound-pressure transducers because frequency range is limited. However, dynamic microphones are quite popular for voice use, since they are self-generating and frequency range limitations present no problem. Note that a loudspeaker is a dynamic microphone of sorts; that is, sound waves striking the cone of a loudspeaker will produce an output signal across the voice coil.

Piezoelectric-type sound transducers. Figure 7-40 shows the construction of a piezoelectric-type sound transducer. This is a form of "crystal microphone" and is also a self-generating device requiring no excitation voltage. The diaphragm is connected to a plate that compresses a piezoelectric material (typically quartz crystal) against a backplate. Sound waves striking the diaphragm compress the crystal and generate an a-c signal that is proportional to the frequency and amplitude of the sound waves. Crystal microphones are quite popular since they have a good frequency range (compared to dynamic microphones) and are self-generating. However, all piezoelectric-type devices are sensitive to vibration. Crystal microphones are also more delicate and more affected by moisture. Piezoelectric-type sound-pressure transducers are often sealed to prevent the entry of moisture, which makes them subject to error at high altitudes or where measurement must be made over a wide range of altitudes (where the reference pressure changes with altitude).

Reluctive-type sound transducers. The reluctive transduction method is not in general use for sound measurement. When used, it generally takes the form of a differential transformer, where a core is attached to a sound-

156

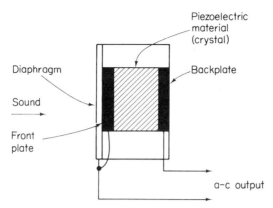

Figure 7-40   Piezoelectric-type sound transducer.

sensing diaphragm, thus requiring an a-c excitation voltage. Also, the frequency response of reluctive transducers is generally inferior to that of other sound transducers.

MEASUREMENTS IN CONTROL AND INSTRUMENTATION SYSTEMS

As can be seen from a review of Chapters 1 through 7, measurement is a very important function of control and instrumentation systems. For example, in a typical industrial control system, it is necessary to measure temperature and pressure of a process variable, the rate of flow, the acidity, and much more. Transducers convert these measurements into electrical quantities such as current, voltage, resistance, etc. In turn, the electrical quantities are displayed as readouts, sometimes as current, voltage, etc., and in other cases in terms of the quantity being measured (temperature, pressure, etc.). The electrical quantities are also used as control signals to operate system controllers or other control devices.

The measurements are needed, not only to learn the condition or quantity of the variable, but also to obtain a value which may be compared with a standard value (the set-point value, discussed in Chapter 1) to obtain an error signal. In a typical control system, this error signal is generated by a controller, and is used to operate an actuator which causes the process variable to return to a predetermined value. In this chapter, we consider the most commonly used methods generally available to measure these control and error signals.

As discussed in Chapters 1 through 7, transducers most often produce signals in the form of changing current, voltage and resistance, or possibly changing capacitance and inductance. Current, voltage, and resistance are generally measured by means of meters, either analog or digital. Sections 8-1 and 8-2 describe the basics of analog and digital meters, respectively, as they relate to instrumentation systems. Capacitance and inductance are usually measured by some form of bridge circuit. Likewise, many transducers are used in a bridge network, and some meters include a bridge network in their circuits. Section 8-3 is devoted to the basics of bridge-type measurement devices.

Many control and instrumentation systems require a count or timing function. For example, the radiation sensors discussed in Chapter 5 require that the pulses produced by radiation be counted over a given period of time. That is, the pulse rate must be measured and displayed. This is the function of a counter. Counters can be mechanical, electrical, or electronic. All three types are discussed in Section 8-4.

8-1 ANALOG METER BASICS

Most meters used in control and instrumentation systems are analog meters. That is, the meters use rectifiers, amplifiers, and other circuits to generate a current proportional to the quantity being measured. The current, in turn, drives a meter movement. Even the digital meters described in Sec. 8-2 are of the analog type, but they differ from the conventional analog meter in the manner of readout. The following paragraphs describe the basic principles of analog meters and show how these basic principles are adapted to control and instrumentation systems.

The simplest and most common instrument that can measure the three basic electrical values (voltage, current, and resistance) is the *volt-ohm-milliammeter,* or VOM, shown in Fig. 8-1. Sometimes, the term *multimeter* is used instead of VOM. Although a VOM is not generally used directly in a control and instrumentation system, a VOM is frequently used to troubleshoot systems. The first improvement on the VOM was the vacuum-tube voltmeter, or VTVM. Today, the VTVM has generally been replaced by electronic voltmeters such as the transistorized meter and the field-effect meter. The sensitivity of an electronic meter is much greater than that of the VOM, since the electronic meter contains a transistor amplifier (Chapter 7) between the meter movement and the input. An electronic meter has another advantage over a VOM in that the electronic meter presents a high impedance to the circuit or component being measured. Thus, an electronic meter draws little or no current from the circuit being tested and has little effect on circuit operation.

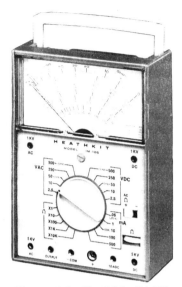

Figure 8-1 Heathkit IM-105 moving-needle VOM. (Courtesy of Heath Co.)

8-1.1 D'Arsonval Movement and the Basic VOM

Except for digital meters (Sec. 8-2) all meter circuits are designed around a basic meter movement. Most nondigital meters use some form of the D'Arsonval meter movement shown in Fig. 8-2. This movement is also known as the *moving-coil galvanometer.* Early D'Arsonval movements had a core made of soft iron. A coil of very fine wire was wound on an aluminum bobbin formed around the core. The iron core is now usually omitted from the movement. The coil and aluminum form function somewhat like an armature mounted on a shaft seated in jewel bearings in order to be free to turn (rotate). Springs on each end of the shaft act as current leads to the coil and help steady the movement.

The coil is placed between the poles of a U-shaped permanent magnet. One end of a pointer is fastened to the armature shaft. As the shaft rotates, the other end of a pointer moves over a calibrated scale. This scale can be calibrated in term of voltage, current, resistance, or of such values as pressure, speed, temperature, etc. Current through the armature coil sets up a magnetic field that reacts with the permanent magnet's field to rotate the coil with respect to the magnet. When current passes through the coil, the magnetic fields are such that the poles repel and, since the permanent magnet cannot move, the coil rotates on the shaft. Current through the coil makes the coil turn a proportional amount. Thus, the basic meter movement is an analog device.

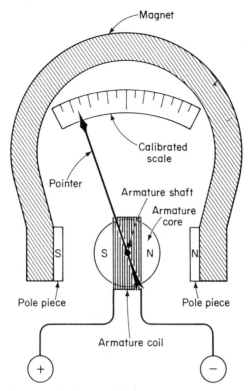

Figure 8-2 Basic D'Arsonval meter movement.

The amount of travel of the pointer attached to the coil is related directly to the amount of current flowing through the movement. The meter scale is then related to some particular current. For example, if 1 mA is required to rotate the coil and pointer across the full scale, a half-scale reading will be equal to 0.5 mA, a quarter-scale reading will be equal to 0.25 mA, etc.

Usually, maximum rotation of the armature (full-scale reading) is completed in less than a half-turn in the clockwise direction. The complete assembly is enclosed in a glass-faced case that protects it from dust and air currents. This enclosed meter movement can be used by itself as a very sensitive *ammeter*. However, the movement is usually part of another instrument, such as a VOM, or of a control panel used in the instrumentation system. Often there is a resistor network to extend the range of the basic movement (as an ammeter), or to convert the basic movement to a voltmeter.

8-1.2 Basic Ammeter

By itself, a basic D'Arsonval movement forms an ammeter (*am*pere *meter*). A true ammeter measures current in amperes. In control and instrumentation systems, current is often measured in milliamperes or

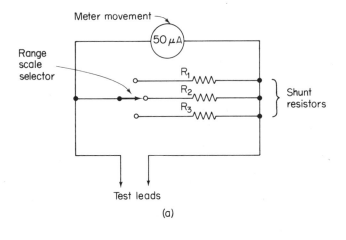

(a)

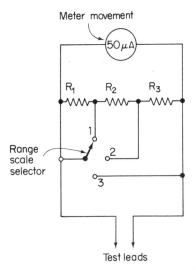

Figure 8-3 Typical milliammeter range selection circuits.

microamperes. A typical movement will produce full-scale deflection when 50 μA of current is passed through the movement.

A *shunt* must be connected across the meter movement if it is desired to measure current greater than the full-scale range of the basic meter. The shunt can be a precision resistor, a bar of metal, or a simple piece of wire. Electronic test meter shunts are usually precision resistors that may be selected by means of a switch. Panel meters for control and instrumentation systems use metal bar shunts. Shunt resistance is only a fraction of the movement resistance. Current divides itself between the meter and shunt, with most of the current flowing through the shunt. Shunts must be precisely calibrated to match the meter movement.

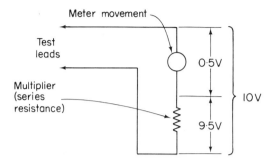

Figure 8-4 Basic voltmeter circuit.

Figure 8-3 shows two typical milliammeter range selection circuits. In Fig. 8-3(a), individual shunts are selected by the range scale selector. In Fig. 8-3(b), the shunts are cut in or out of the circuit by the selector. If the selector is in position 1, all three shunts are across the meter movement, giving the least shunting effect (most current through the movement). In position 2, resistor $R1$ is shorted out of the circuit, with resistors $R2$ and $R3$ shunted across the movement, increasing the meter's current range. In position 3, only $R3$ is shunted across the movement, and the meter reads maximum current.

8-1.3 Basic Voltmeter

When a basic D'Arsonval movement is connected in series with resistors, a voltmeter is formed. The series resistance is known as a *multiplier*, since the resistance multiplies the range of the basic meter movement. The basic voltmeter circuit is shown in Fig. 8-4. As shown, the voltage divides itself across the meter movement and the series resistance. If an 0.5 V full-scale deflection meter movement is used and it is desired to measure a full scale of 10 V, the series resistor must drop 9.5 V. If a 100 V full scale is desired, the series resistance must drop 99.5 V, etc.

Figure 8-5 shows two typical voltmeter range selection circuits. In Fig. 8-5(a), individual multipliers are selected by the range-scale selector. In Fig. 8-5(b), the multipliers are cut in or out of the circuit by the selector. If the selector is in position 1, only resistor $R1$ is in the circuit, giving the least voltage drop (meter will read the lowest range). In position 2, both $R1$ and $R2$ are in the circuit, giving the meter a higher voltage range. In position 3, all three resistors drop the voltage, permitting the meter to read maximum voltage.

The term *ohms per volt* is used to describe meters. Ohms per volt is a measure of a meter's sensitivity and represents the number of ohms required to extend the range by 1 V. For example, if the meter movement requires 1 mA for full-scale deflection, the 1000 ohms (including the movement's internal resistance) are needed for each volt that could be measured. If the

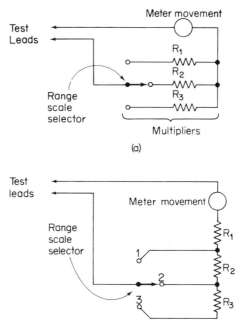

Figure 8-5 Typical voltmeter range selection circuits.

movement requires only 50 μA for full-scale deflection, then 20,000 ohms per volt are needed. Thus, the more sensitive the meter movement (those requiring the least current), the higher the ohms per volt requirement. Voltmeters with a high ohms per volt rating put less load on (draw less current from) the circuit being measured and have a less disturbing effect on the circuit.

8-1.4 Basic Ohmmeter

An *ohmmeter* (or resistance-measuring device) is formed when a basic meter movement is connected in series with a resistance and a power source (such as a battery in portable meters). The basic ohmmeter arrangement is shown in Fig. 8-6. Here, a 3 V battery is connected to a meter movement with a full-scale reading of 50 μA. The current-limiting resistor R has a value (60 kilohms, less meter resistance) such that exactly 50 μA will flow in the circuit when the test leads are clipped together.

When there is no connection across the test leads, the current is zero. The meter's pointer rests at the "infinity" mark (∞) on the scale. When the two leads are shorted, the meter moves to the full 50 uA reading which, on the scale, indicates a "zero ohms" reading. If a 60 kilohm resistance is connected across the leads, as shown in Fig. 8-6, the total resistance is 120

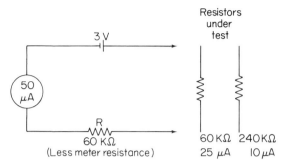

Figure 8-6 Basic ohmmeter circuit.

kilohms, and the meter drops to one-half of full-scale reading, or 25 μA. If the battery voltage and limiting resistor R remain constant, the pointer will always move to 25 μA whenever 60 kilohms are connected across the test leads. The 25 μA point on the meter can then be marked "60 kilohms."

With a 240 kilohm resistance across the leads, the total resistance is 300 kilohms, and the pointer drops to a 10 μA reading, since $I = E/R$, or $3/300,000 = 10$ μA. Again, if the battery voltage and the limiting resistor remain constant, the meter will always read 10 μA when a resistance of 240 kilohms is placed across the test leads. Thus, the 10 μA point on the meter scale can be marked "240 kilohms."

The ohmmeter circuit of Fig. 8-6 is, therefore, capable of measuring 60 and 240 kilohms. Any number of resistance values can be plotted on the scale, provided resistances of known value are placed across the leads.

As shown in Fig. 8-1, the ohmmeter scale of a typical VOM is printed on the meter face along with the voltage and current scales. However, the ohmmeter scale is quite different from the other scales in two respects. The zero point is at the right-hand side (usually), and the maximum resistance (usually marked "infinity") is at the left-hand side. Also, the scale is not linear (lower-resistance divisions are wider and higher-resistance divisions are narrower).

The ohmmeter circuit of a typical VOM is shown in Fig. 8-7. Here, the ohmmeter has five range scales that can be selected by means of a switch. In all but the "X1" position, a series multiplier (similar to that of a voltmeter) is connected to the circuit and drops the voltage by a corresponding amount. This reduces current flow through the entire circuit, usually by a ratio of 10:1, 100:1, 1000:1, or 10,000:1 so that the ohmmeter scale represents 10, 100, 1000, or 10,000 times the indicated amount.

No matter which scale is used, the meter and battery are in series with a variable resistor that allows the circuit to be "zeroed." As a battery ages, the output voltage drops. Also, it is possible that with extended age or extreme temperature, the resistance values (or meter movement itself) could

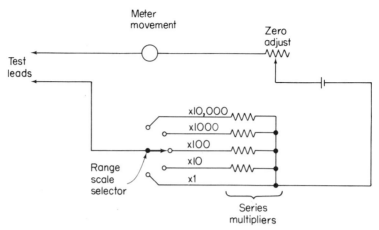

Figure 8-7 Ohmmeter circuit of a typical VOM.

change in value. Any of these conditions can make the ohmmeter scale inaccurate. The variable resistor (usually marked "zero adjust" or "zero") compensates for these problems. In use, the test leads are shorted together, and the variable resistor is adjusted until the meter is at zero (at the right-hand side of the ohmmeter scale). When the leads are opened, the meter pointer then drops back to "infinity" or "open" (left-hand side), and the meter is ready to read resistance accurately.

8-1.5 Basic Galvanometer

The basic meter movement described thus far can be used as a galvanometer. However, the term *galvanometer* has come to mean a meter where the zero of the scale is at the center, with negative current reading to the left and positive current reading to the right as shown in Fig. 8-8. Generally, a galvanometer is used to read proportional positive or negative changes in circuits rather than the actual unit value of current. In a control and instrumentation system, a galvanometer can be used to indicate some quantity such as pressure, temperature, etc., above or below a given level. For such applications, the scale is calibrated in terms of the quantity being measured (pressure, temperature, etc.). However, the main use for such a meter is in bridge circuits, such as those described in Sec. 8-3.

8-1.6 Basic a-c Meters

Most a-c meters are similar to d-c meters in that they are both analog current-measuring devices. However, since a-c reverses direction during each cycle, the basic meter movement cannot be connected directly to a-c. Instead, the meter movement is connected to the a-c voltage through a rectifier (Chapter 7). Both half-wave and full-wave rectifiers are used. However, the full-wave bridge rectifier of Fig. 8-9 is most efficient, since a direct current will flow through the meter movement on both half-cycles. The remainder of the a-c meter circuit can be identical to that of a d-c meter.

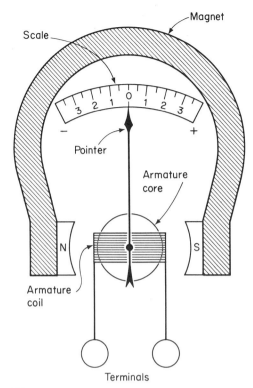

Figure 8-8 Basic zero-center galvanometer movement.

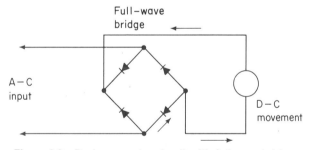

Figure 8-9 Basic a-c meter circuit with full-wave bridge.

Alternating-current meter scales present a problem that does not occur for d-c scales. As shown in Fig. 8-10, there are four ways to measure an a-c voltage. We can measure the *average,* RMS or *effective, peak,* or *peak-to-peak* voltage.

Peak voltage is measured from the crest of one half-cycle, whereas *peak-to-peak* is measured from the crests of both half-cycles. However, the direct current to the meter movement is less than the peak alternating current, since the voltage and current drop to zero on each half-cycle.

With a full-wave bridge rectifier, the current or voltage is 0.637 of the peak value (a half-wave rectifier delivers 0.318 of the peak value). This is

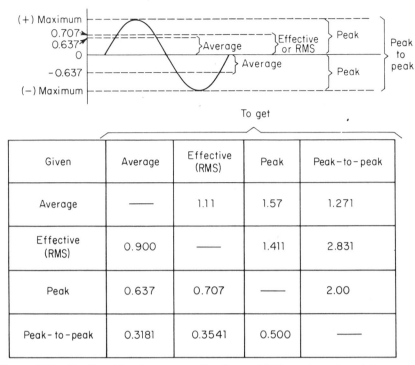

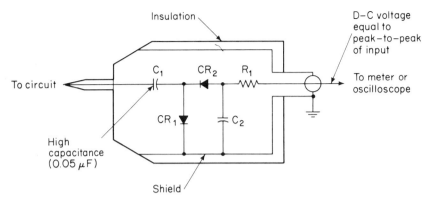

Figure 8-11 Typical full-wave radio-frequency probe circuit.

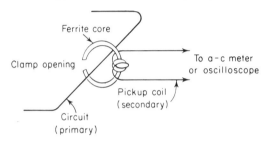

Figure 8-12 Basic clip-on meter or current probe circuit.

d-c output voltage that is equal to the peak r-f voltage. The d-c output of the probe is then applied to the meter and is displayed as a voltage readout in the normal manner.

A full-wave r-f probe is shown in Fig. 8-11. This circuit can be used to provide either peak-to-peak output or RMS output. Capacitor $C1$ is a high-capacitance d-c blocking capacitor used to protect diode $CR1$. Usually, a germanium diode is used for $CR1$, which rectifies the r-f voltage and produces a d-c output across $R1$. For some probes, $R1$ is omitted so that the d-c voltage is developed directly across the input circuit of the meter. That d-c voltage is equal to the peak r-f voltage, less whatever voltage drop exists across the diode $CR1$. Since some meters are calibrated in RMS, the probe output is reduced to 0.3535 of the peak-to-peak value, accomplished by selecting or adjusting the value of $R1$.

8-1.7 Current Probes and Clip-On Meters

One type of meter that is unique to a-c measurements is the clip-on meter or current probe shown in Fig. 8-12. Alternating currents set up alternating fields around a wire or conductor. These currents can be picked up by a coil of wire around the conductor, stepped up through a transformer, and measured by a voltmeter. A clip-on meter, complete with built-in coil, transformer, and meter movement, is particularly useful where conductors are carrying heavy currents and where it is not convenient to insert an ammeter in the line. Clip-on meters are most often used for heavy industrial control and instrumentation systems.

Current probes are similar to clip-on meters, except that probes are generally used in conjunction with an amplifier to measure small currents. Most electronic laboratory instrumentation systems use current probes rather than clip-on meters. A typical probe clips around the wire carrying the current to be measured and, in effect, makes the wire the one-turn primary of a transformer formed by ferrite cores and a many-turned secondary within the probe.

8-1.8 Thermocouple Meters

Figure 8-13 shows the basic circuit used in thermocouple meters. Such meters measure d-c, a-c, and even r-f currents. As mentioned in Chapter 6, when two dissimilar metals are connected at one end and heat is applied to the connected ends, a d-c voltage is developed across the open ends of the two dissimilar metals. This voltage is directly proportional to the temperature of the wires in the heated junction of the thermocouple. An electric current passing through a wire or conductor produces heat in that wire in proportion to the square of the current. Thus, if a current is passed through the junction of a thermocouple, heat is generated in the wires, and a voltage is developed at the open ends.

If a calibrated meter movement is connected to the free ends of the thermocouple as shown in Fig. 8-13, the generated voltage can be measured. Of

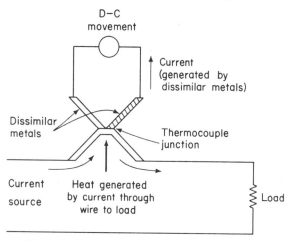

Figure 8-13 Basic thermocouple meter circuit.

course, the meter scale must be calibrated to relate the reading to the amount of current passing through the thermocouple and heating the junction, rather than the voltage produced by the thermocouple. The direction of current in the thermocouple has no effect on the heating of the junction, so the instrument can measure a-c, d-c, or r-f current. For measuring very low currents or very high frequencies, the thermocouple junction is usually sealed in a vacuum similar to the filament of a vacuum tube, thus providing the greatest amount of heat for a minimum amount of current.

8-1.9 Laboratory Analog Meters

The meters used for laboratory instrumentation systems operate by producing a current proportional to the quantity being measured, as do basic analog meters. However, laboratory instrumentation meters include many circuit refinements to improve their accuracy and stability. Also, there are special-purpose analog meter circuits that are unique to laboratory instruments.

Basic electronic meter measurements. The basic electronic meter circuit is shown in Fig. 8-14. The amplifier (shown as a triangle symbol) uses transistors (Chapter 7), often field-effect transistors, or FET's, as discussed in Sec. 7-3.3.

When the basic circuit is used as a voltmeter, resistance R has a large value (usually several megohms) and is connected in parallel across the voltage circuit being measured. Because of the high resistance, very little current is drawn from the circuit, and operation of the circuit is not disturbed. The voltage across resistance R raises the voltage at the amplifier input from the zero level. This causes the meter at the amplifier output to indicate a corresponding voltage.

When the basic circuit is used as an ammeter, resistance R has a small value (a few ohms or less) and is connected in series with the circuit being measured. Because of the low resistance, there is little change in the total circuit current, and operation of the circuit is not disturbed. Current flow through resistance R causes a voltage to be developed across R, which raises the amplifier input level from zero and causes the meter to indicate a corresponding (current) reading.

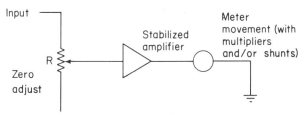

Figure 8-14   Basic electronic voltmeter circuit.

One of the most common circuits used in electronic meters is shown in Fig. 8-15. This circuit is essentially a *differential amplifier* (Sec. 8-2) in which the voltage to be measured is applied to one input and the other input is grounded. The zero-set resistance is adjusted so that the meter reads zero when no input voltage is applied. When the voltage to be measured is applied across the input resistance, the circuit is unbalanced, and the meter indicates the proportional unbalance as a corresponding voltage reading. One of the reasons for using the differential amplifier circuit is to minimize drift due to power supply changes (as discussed in the following paragraphs).

The amplifiers shown in Figs. 8-14 and 8-15 perform three basic functions. First, the effective sensitivity of the meter movement is increased. An amplifier changes the measured quantity to a current of sufficient amplitude to deflect the meter movement. Thus, a few microvolts that would not show up on a typical meter movement could be amplified to several volts (which is sufficient to deflect any meter movement).

The second function of the amplifier is to increase the input impedance of the meter so that the instrument draws little current from the circuit under test. A typical FET meter amplifier can provide up to about 100 megohms of impedance.

The third amplifier function is to limit the maximum current applied to the meter movement. Therefore, there is little danger when unexpected overloads occur that could burn out the meter movement.

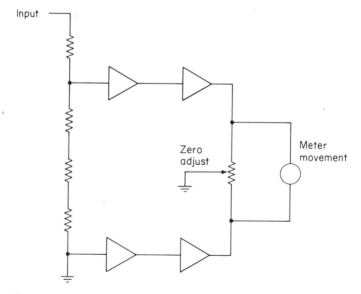

Figure 8-15 Basic electronic voltmeter with differential amplifier.

Drift problems of electronic meters. One of the problems common to an amplifier is drift due to power supply changes. The amplifier cannot tell the difference between power supply change and a change in the voltage being measured. This problem is especially aggravated when an electronic meter is used to measure small voltages. A common technique for eliminating drift is to convert the d-c voltage being measured into an a-c voltage. One way to make the conversion is by use of a d-c to a-c converter described in Sec. 7-3.4. A more common method for meter instrumentation applications is to alternately apply and remove the d-c input voltage through a "chopper," amplify the "chopped" signal in an a-c amplifier, and then synchronously rectify the amplifier output back to a d-c voltage for application on a basic meter movement.

An electromechanical chopper can be used, but some form of electronic chopper (such as the photoconductive chopper shown in Fig. 8-16) is more common. In a typical photoconductive chopper, the d-c input voltage is converted to a comparable a-c voltage by periodically illuminating a group of photoconductive light sensors or resistors (Sec. 5-1.4). The photoconductive resistors are illuminated by a flasher such as a neon lamp driven by an a-c voltage source. The input and output signals are synchronized, since

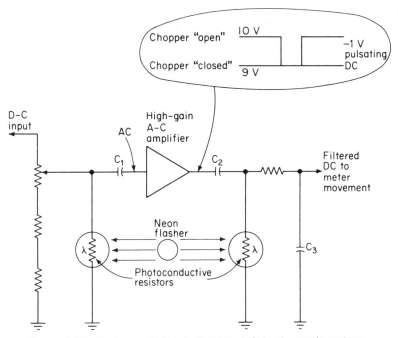

Figure 8-16 Basic circuit for eliminating drift in electronic meters.

they are both driven by the same source. When the photoconductive resistors are illuminated they provide low resistance and act as a short circuit. Without light, the resistors provide very high resistance and appear as an open circuit.

With light, the input to the amplifier is grounded (zero input). One side of capacitor $C2$ is also grounded. The opposite side of $C2$ is connected to the amplifier output circuit. Capacitor $C2$ charges up to the output value. When the flasher light is removed on alternate half-cycles of the driving voltage, the d-c voltage to be measured is applied to the amplifier input, causing the amplifier output to drop by an amount proportional to the input signal. Capacitor $C2$ then discharges into the meter circuit by a corresponding amount. For example, assume that $C2$ is charged to 10 V with no input and that the amplifier output drops to 9 V when the input is applied (chopper open). Under these conditions, the output is then ideally a square wave equal to $-1V$.

Resistance measurements in electronic meters.   In a VOM or multimeter, resistance is measured by applying a known voltage to an unknown resistance and then measuring the current passing through the circuit. When voltage and current are known, resistance can be computed. In actual practice, computation is unnecessary since the resistance scale of the meter is precalculated.

Most electronic meters use a modified procedure for resistance measurement. As shown in Fig. 8-17, the current in the circuit depends on the series combination of the unknown resistor Rx and the internal resistor Ri. This means that both the voltage and current in the external circuit will change according to the value of the unknown resistance. If Rx is infinite, the meter reads the full battery voltage. Full-scale deflection corresponds to a resistance of infinity. If Rx is zero (short circuit), the meter reads zero. The mid-scale range occurs when Rx equals Ri.

The resistance Ri included as part of the ohmmeter circuit provides a convenient means of changing the range of the instrument. When values of low resistance are being measured, the resistance of the ohmmeter leads (in-

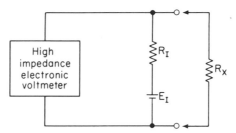

Figure 8-17 Basic resistance measurement procedures in electronic meters.

cluded in the total resistance measurement) can cause a considerable error To overcome this problem, the circuit can be altered to that shown in Fig. 8-18. Here, the resistance of the current-carrying leads is calibrated as part of R_i, and the resistance in the voltmeter leads is small compared with the high input impedance of the metering circuit.

Alternating current measurements in electronic meters. An electronic meter for measuring a-c voltages also uses an amplifier with the meter movement but adds a rectifier circuit to convert the a-c to d-c. Most a-c meters are RMS-reading instruments, although it is possible for the scales to read average, peak, or peak-to-peak. Although a meter may be RMS-reading, it is usually average or peak-responding. That is, the scale reads RMS values, but the meter movement operates on an average or peak value.

Figure 8-19 shows the basic circuit of an *average-responding meter.* Here, the a-c signal is amplified and rectified. The resulting current pulses drive the meter movement. Meter deflection is proportional to the average value of the waveform being measured (even though the scale may or may not read RMS).

The *peak-responding voltmeter* shown in Fig. 8-20 places the rectifier in the input circuit, where a small capacitor is charged to the peak value of the input voltage. This voltage is amplified and passed to the meter movement. Meter deflection is proportional to the peak amplitude of the input waveform. The meter scale can be calibrated in RMS or peak voltage, as required.

If highly complex waveforms (non-sine waves) are to be measured, a true RMS-responding voltmeter should be used. Such a circuit is shown in Fig. 8-21. Here, the complex waveform is used to heat the junction of thermocouples. One of the major problems of this technique is the nonlinear behavior, as well as low response and possible burnout, of the thermocouple. Nonlinear behavior complicates calibration of the indicating meter. This difficulty can be overcome by the use of two thermocouples mounted in the same thermal environment. Nonlinear effects in the measuring ther-

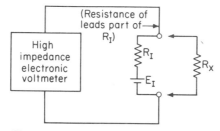

Figure 8-18 Improved resistance measurement procedure for electronic voltmeters.

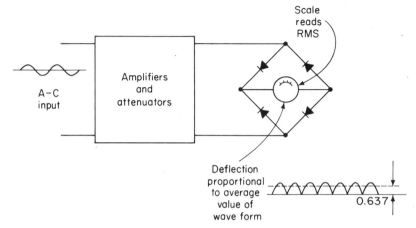

Figure 8-19 Basic average-responding (RMS-reading) a-c voltmeter.

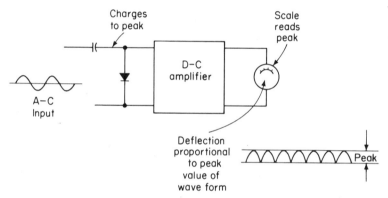

Figure 8-20 Basic peak-responding a-c voltmeter.

mocouple are offset by similar nonlinear operations of the second thermocouple.

As shown in the circuit of Fig. 8-21, developed by Hewlett-Packard, the amplified input signal is applied to the measuring thermocouple, and a d-c feedback voltage is fed to the balancing thermocouple. The d-c voltage is obtained from the voltage output difference between the thermocouples. The circuit can be considered as a feedback control system that matches the heating power of the d-c feedback voltage to the input waveform heating power. Meter deflection is proportional to the d-c feedback voltage which, in turn, is proportional to the RMS of the input voltage (no matter what the waveform). Thus, the meter indication is linear.

It is also possible to measure peak-to-peak voltages with an electronic voltmeter. The circuit is similar to that of the peak-responding circuit of Fig. 8-20. However, the input capacitor is charged to the peak-to-peak value

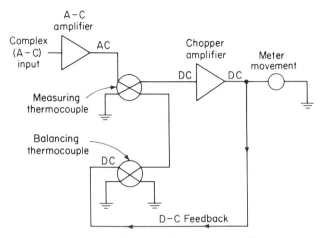

Figure 8-21 True RMS-responding voltmeter for measurement of complex waves.

by a full-wave rectifier. Also, the scale is calibrated to read directly in peak-to-peak values.

8-1.10   Power Measurement

There are several methods used to measure power in control and instrumentation systems. Although the methods are relatively simple, it is necessary to know the characteristics of power in electrical circuits to understand how the measurements are made.

The basic method for measuring the electrical power consumed by a load is to connect an ammeter in series with the load, and a voltmeter across the load. Then, since the power (in *watts*) is equal to the product of the current flowing through the load (in *amperes*), and the voltage (in *volts*) applied across the load, the power can be determined by multiplying the readings of the two meters.

The power, in watts, is thus indicated in d-c circuits (where there is no *phase difference* between current and voltage), or in a-c circuits where only a resistive load is concerned, and the current and voltage are in phase. In an a-c circuit, where the load has an *inductive* or *capacitive* component, a new factor, called *reactance,* is introduced. As a result, the current and voltage of the circuit no longer are in phase, and the phase difference is a function of the amount of inductance or capacitance (or both).

There are two kinds of power in an a-c circuit with an inductive or capacitive component. One is *true* or useful power that can do useful work (such as causing a motor to rotate). The other power is useless, or *reactive,* power that can do no useful work. Total power, which is known as *apparent power,* is the sum of true power and reactive power. The ratio between true

177

power and apparent power in an a-c circuit is known as the *power factor* of the circuit.

Power factor equals true power divided by apparent power. Likewise, true power equals apparent power multiplied by power factor. This power factor, which is always less than 100%, is a function of the phase difference between the current and voltage in a circuit. The apparent power in a circuit may be measured by the basic voltmeter/ammeter method. However, to distinguish apparent power from true power, apparent power is stated in units of *volt-amperes.* True power, which is stated in units of *watts,* is measured by a wattmeter.

Figure 8-22 shows the circuit of a wattmeter used to measure power in 60 Hz power line circuits. As shown, the instrument consists of fixed and moving coils. The two fixed coils, with few turns of heavy wire, are connected in series with the line, and their magnetic fields are a function of the current in the circuit. The moving coil, with many turns of fine wire, together with a series multiplier resistor *R,* is connected in parallel with the load, and the magnetic field is a function of the voltage across the load. Rotation of the moving coil, and the pointer attached to the coil, is a function of the product of the instantaneous values of current and voltage in the circuit. The scale is calibrated to indicate that product in watts. Similar circuits are available to measure power in control systems using three-phase power. A separate set of fixed and moving coils is used for each phase. The scale pointer is connected to all moving coils, and movement of the pointer is determined by the power in all three phases. Thus, the scale over which the pointer moves is calibrated to indicate the total true power of the entire system.

Figure 8-23 shows the circuit of a wattmeter used to measure power in higher frequency systems (such as the output of a radio transmitter). Here, the output of the circuit is applied to a fixed load (often called a *dummy*

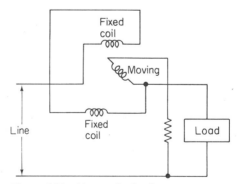

Figure 8-22 How a single-phase wattmeter is connected to measure the true power consumed by a load.

load). Typically, the load is a noninductive (not wire wound) carbon or composition resistor. The value of this load resistor is approximately equal to the output impedance of the circuit being measured. The meter, which is actually an a-c voltmeter (possibly a thermocouple type), has a scale calibrated in watts.

8-2 DIGITAL AND DIFFERENTIAL METER BASICS

To fully understand operation of digital meters, or any digital instrument used in control and instrumentation systems, it is necessary to have a sound understanding of digital logic circuits. These include gates, amplifiers, switching elements, delay elements, binary counting systems, truth tables, registers, encoders, decoders, D/A converters, A/D converters, adders, scalers, etc. Full descriptions of these devices are contained in the author's *Logic Designer's Manual* referenced in Sec. 7-3.5. However, it is possible to have an adequate understanding of digital systems if you can follow simplified block diagrams as presented here.

In actual practice, most digital equipment (meters, counters, and even computer controllers) is comprised of "logic building blocks" (gates, registers, etc.) that are interconnected to perform various functions (mathematical operations, conversion, readout, etc.). If you understand operation of the individual building blocks, you can then understand operation of the complete instrument. Thus, operating principles of complex digital instrumentation systems (including meters, counters, etc.) can be presented in block form, or in logic-diagram form, with blocks and logic symbols representing the building blocks. This method of presentation is followed by most manufacturers of digital instrumentation in their instruction manuals.

It should also be noted that a knowledge of *electronic counters* is necessary to fully understand digital meters because a digital meter performs two basic functions: (1) conversion of voltage (or other quantity being measured) to time or frequency and (2) conversion of the time or frequency data to a digital readout. In effect, a digital meter is a conversion

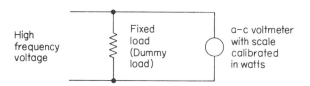

Figure 8-23 Wattmeter used to measure power in higher frequency systems.

circuit (voltage-to-time, etc.) plus an electronic counter for readout. Operation of electronic counters is discussed in Sec. 8-4.

8-2.1 Basic Digital Meter

Digital instruments are available to measure a-c and d-c voltages, currents, and resistance. Other physical variables can also be measured by use of suitable transducers, as described in Chapters 1 through 7. Many digital instruments have outputs that can be used to make permanent records of measurements with printers, cards and tape punches, and magnetic tape equipment. With data in digital form, it may be processed by computers and computer controllers with no loss of accuracy.

The most popular digital instrument is the *digital multimeter* such as shown in Fig. 8-24. Such instruments display measurements as *discrete numerals,* rather than as a pointer deflection on a continuous scale commonly used in analog meters described in Sec. 8-1. Direct numerical readout reduces human error, eliminates parallax error (which is an error in observation that occurs when the operator's eye is not directly over the meter pointer), and increases reading speed. Automatic polarity and range-changing features on most digital instruments reduce operator training, measurement error, and possible instrument damage through overload.

Note the simplicity of operating controls shown in Fig. 8-24. Once the power is turned on, the operator has only to select the desired range and type of measurement (a-c, d-c, voltage, current, resistance). The meter can then be connected to the circuit and the readout will be automatic. On some digital instruments, the range is changed automatically, further simplifying operation.

There are many types of digital meters. We discuss only one type here, the *ramp type.*

Figure 8-24 Heathkit IM-2202 digital multimeter. (Courtesy of Heath Co.)

8-2.2 Ramp-Type Digital Meter

Figure 8-25 shows the block diagram of a typical ramp-type digital voltmeter. The operating principle of this instrument is to measure the time required for a linear ramp to change from a value equal to the voltage being measured to zero (or vice versa). The time period is measured with an electronic time-interval counter (Sec. 8-4) and is displayed on a *decade readout.* As discussed in Chapter 10, with a decade readout *each digit* in the total readout is capable of being displayed as numerals running from 0 through 9. The readout of Fig. 8-24 is capable of displaying any combination of numerals from 0000 to 9999, whereas the readout of Fig. 8-25 can display combinations of numerals from 00000 to 99999, plus a polarity digit.

The ramp-type meter is essentially a voltage-to-time converter, plus a counter and readout. The timing diagram of Fig. 8-26 illustrates conversion of a voltage to a time interval. At the start of a measurement cycle (there are usually two or three measurement cycles per second) a ramp voltage is generated. The ramp voltage is compared continuously with the voltage being measured. At the instant the two voltages become equal, a coincidence circuit generates a pulse that opens a gate. The ramp continues until a second comparator circuit senses that the ramp has reached 0 V. The output pulse of the comparator closes the gate. The time duration of the gate opening is proportional to the input voltage. The gate allows a pulse to pass to the counter circuit, and the number of pulses counted during the gating interval is a measure of the voltage.

The elapsed time, as indicated by the count, is proportional to the time

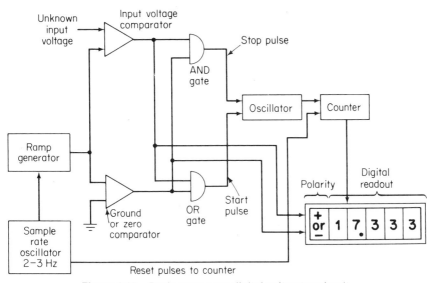

Figure 8-25 Basic ramp-type digital voltmeter circuit.

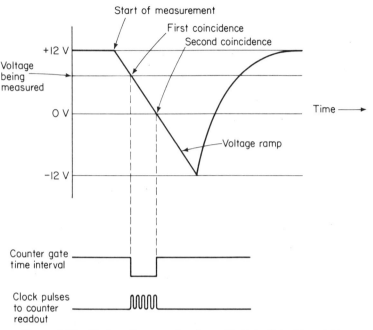

Figure 8-26 Timing diagram showing voltage-to-time conversion.

the ramp takes to travel between the unknown voltage and 0 V (or vice versa, in the case of a negative input voltage to be measured). Therefore, the count is equal to the input voltage. The order in which pulses come from the two comparators indicates the polarity of the unknown voltage, triggering a readout that indicates plus or minus, as required.

Staircase ramp digital meter. Figure 8-27 is a block diagram of a typical staircase ramp instrument that is an improvement over the basic ramp type. The meter of Fig. 8-27 makes voltage measurements by comparing the input voltage to an internally generated staircase ramp voltage, rather than to a linear ramp. When the input and the staircase ramp voltages are equal, the comparator generates a signal to stop the ramp and the count. The instrument then displays the number of counts that are necessary to make the staircase ramp equal to the input voltage. At the end of the sample (the sample rate is fixed at two samples per second by the 2 Hz sample oscillator) a reset pulse resets the staircase ramp to zero, and the measurement starts over again.

8-2.3 Basic Differential Meter

Figure 8-28 shows a basic differential voltage measurement circuit. Note that the null meter shown is a form of galvanometer such as discussed in Sec. 8-1.5. The basic concept of differential voltage measurement is to ap-

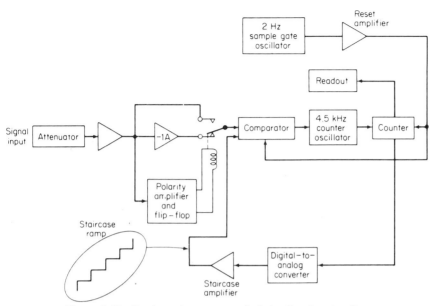

Figure 8-27 Basic staircase-ramp digital voltmeter circuit.

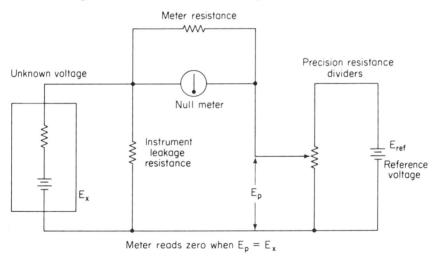

Meter reads zero when $E_p = E_x$

Figure 8-28 Basic differential voltage measurment.

ply an unknown voltage against one that is accurately known and then to measure the difference between the two on an indicating device. If the known voltage is adjusted to the exact potential of the unknown voltage, one can determine the unknown quantity being measured as accurately as the known voltage (or reference standard).

Measurements made by the differential voltmeter technique (the device is sometimes called a *potentiometric* or *manual voltmeter*) are recognized as one of the most accurate means of relating an unknown voltage to a known reference voltage. As shown in Fig. 8-28, these measurements are made by adjusting a *precision resistive divider* to divide down an accurately known reference voltage. The divider is adjusted to the point where the divider output equals the unknown voltage, as shown by a zero or center-scale reading on the null meter.

The unknown voltage is determined to an accuracy limited only by the accuracies of the reference voltage and the resistive divider. Meter accuracy is of little consequence, since the meter serves only to indicate any residual differential between the known and unknown voltages.

With this system a high-voltage standard is required to measure high voltage. This need may be overcome by inserting a voltage divider between the source and the null meter as shown in Fig. 8-29. This expedient, however, results in relatively low input resistance for voltages higher than the reference standard. A low input resistance is undesirable because accurate measurement may not be obtained if substantial current is drawn from the source being measured. Most differential voltmeters used today offer input resistance approaching infinity only at a null condition, and then only if an input voltage divider is not used.

To overcome these limitations, Hewlett-Packard has developed a differential voltmeter with an input isolation stage. A simplified diagram of the circuit is shown in Fig. 8-30. Isolation is accomplished by means of an amplifier between the measurement source and the measurement circuits. The amplifier ensures that the high input impedance is maintained regardless of whether the instrument is adjusted for a null reading.

A further advantage provided by the amplifier is that the resistive voltage divider that enables voltages as high as 1000 V to be compared to a precision 1 V reference may be placed at the output of the amplifier, rather than being in series with the measured voltage source. This isolation permits the instrument to have high impedance on all ranges.

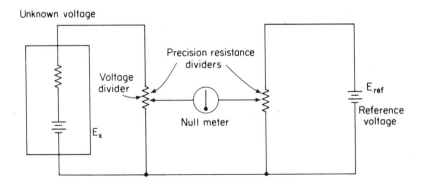

Figure 8-29 Potentiometric method of measuring unknown voltages.

8-3 BRIDGE-TYPE TEST EQUIPMENT

Many quantities such as capacitance, resistance, inductance, etc. are measured by means of bridge circuits and bridge-type instruments. Some manufacturers, such as Hewlett-Packard and General Radio, produce *universal bridges* that will measure more than one quantity. These instruments operate by the *balance* or *null principle,* or on the principle of *comparison against a standard.* This section is devoted to instruments that operate on null or comparison techniques.

8-3.1 Wheatstone Bridges

Most null or balance-type instruments are evolved from the basic Wheatstone bridge shown in Fig. 8-31. Note that this bridge circuit is similar to that used in many transducers. As discussed in Chapters 1 through 7, the transduction element forms one leg of the bridge in such transducers. In the circuit of Fig. 8-31, resistors R_A and R_B are fixed and of known value. R_S is a variable resistor with the necessary calibration arrangement to read the resistance value for any setting (usually a calibrated dial coupled to the variable resistance shaft). The unknown resistance value R_X is connected across terminals B and C. A battery or other power source is connected across points A and C.

When switch $S1$ is closed, current flows in the direction of the arrows, and there is a voltage drop across all four resistors. The drop across R_A is equal to the drop across R_B (provided that R_A and R_B are of equal resistance value). Variable resistance R_S is adjusted so that the galvanometer reads zero (center scale) when switch $S2$ is closed. At this adjustment, R_S is equal to R_X in resistance. By reading the resistance of R_S (from the calibrated dial), the resistance of R_X is known.

8-3.2 A-c Bridges

The Wheatstone bridge circuit is easily adapted to a-c measurements. With complex *impedances,* two balance conditions must be satisfied, one for the *resistive* component of the impedance and one for the *reactive* com-

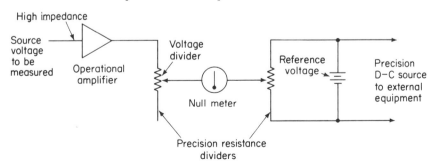

Figure 8-30 Potentiometric voltmeter with high-impedance input amplifier.

185

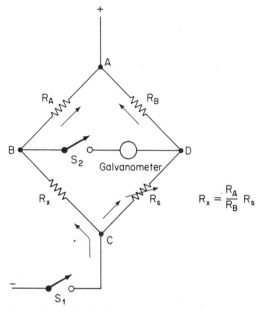

Figure 8-31 Basic Wheatstone resistance bridge.

ponent of the impedance. Likewise, both the *conductive* (or reciprocal of resistive) and *susceptance* (the reciprocal of reactance) must be satisfied for complex *admittances* (the reciprocal of impedances). These balance conditions are shown in Fig. 8-32.

As in the case of d-c bridges, balance for the resistive component in an a-c bridge is obtained by a precision variable resistance. Balance for the reactive component can be obtained from a similar reactance in an adjacent arm of the bridge, or an unlike reactance in the opposite arm. In most practical circuits, the reactance is supplied by means of a fixed, precision capacitor in series or parallel with a variable resistance.

8-3.3 Universal Bridges

A universal bridge is a single instrument that measures capacitance, resistance, and inductance. A generalized universal bridge circuit is shown in Fig. 8-33. The bridge is driven by an a-c source across points O and Q. When the voltage across arm OP equals the voltage across arm OS, the output voltage, expressed across the detector connected to P and S, is zero. With the bridge balanced or nulled, the product of the impedance across OS and that across PQ is equal to the product of the impedance across SQ and that across OP. At balance, the value of any of the four impedances can be calculated if the other three are known.

In addition to the basic bridge circuit, a typical universal bridge contains

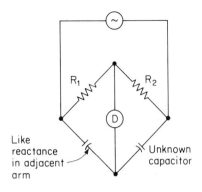

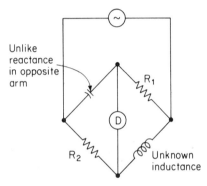

Figure 8-32 Bridge circuits showing balance for reactive component.

an audio generator as an a-c driving source, and a null detector circuit. The null detector includes an amplifier circuit (to measure extremely low voltages at or near null), a fixed or tunable filter or frequency-selective circuit (to remove any signals but the audio driving source), and rectifiers (to convert the a-c signal to a d-c voltage for display on the meter. In some universal briges, the a-c driving source is tunable, whereas in other circuits the driving source is fixed (typically at 1000 Hz). A typical universal bridge also contains a d-c supply (batteries) for measurement of resistance. Many universal bridges are provided with connectors that permit external driving sources and external standards to be used with the internal bridge circuit.

In use, the *DQ* and *CGRL* dials are adjusted for a null (the *DQ* dial is not used for resistance and conductance measurements). Then, the value is read from the corresponding dial indication. The letters *CGRL* stand for capacitance, conductance, resistance, and inductance, respectively. The letters *DQ* stand for dissipation factor and storage factor, respectively. *D* is the ratio of resistance to reactance, and *Q* is its reciprocal. In most cases,

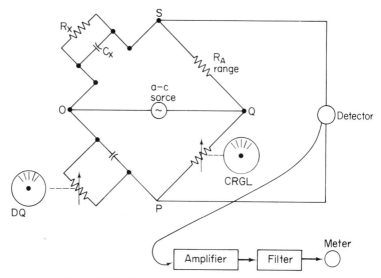

Figure 8-33 Generalized universal bridge circuit.

the dials are not direct reading. Instead, the dial reading must be multiplied by some factor.

In universal bridges, both capacitance and inductance are measured in terms of the impedance they present at a given frequency, rather than by comparison against a standard, even though the readout is in terms of capacity and inductance value (μF, μH, etc.). This is the major difference between the universal bridge circuit and the standard capacitance or inductance bridge described in Sec. 8-3.4.

8-3.4 Standard Capacitance and Inductance Bridges

Capacitance can be measured on a standard capacitance bridge, rather than on a universal bridge. Standard-type bridges provide greater accuracy (generally) than universal bridges but are limited in the variety of measurements that can be made. Standard inductance bridges are also available but are not in such common use as capacitance bridges.

Standard bridges operate on the principle of comparing an unknown against a standard. The basic standard bridge circuit for capacitance measurement is shown in Fig. 8-34. The capacitance of the unknown C_X is balanced by a calibrated, variable, standard capacitor C_N or by a fixed standard capacitor and a variable ratio arm such as R_A. Such bridges with resistive ratio arms and with calibrated variable capacitors or resistors can be used over a wide range of both capacitance and frequency with great accuracy.

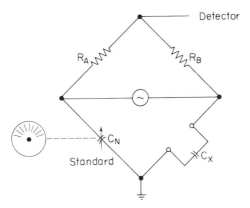

Figure 8-34 Basic standard capacitance bridge circuit.

Figure 8-35 shows the basic circuits of some standard bridges that use the *transformer-ratio* system. In Fig. 8-35(a), the variable standard capacitor is adjusted for a balance indication on the a-c null meter, and the capacitance is read out on the calibrated dial. In Fig. 8-35(b), the standard capacitor is fixed, and the null meter is balanced by means of a variable inductance. In the circuit of Fig. 8-35(c), there are several standard capacitors, each connected to a tap on a selector switch. The standard capacitors are selected, in turn, until a balance is obtained.

8-4 TIMERS

The common unit of time is the *day,* the time it takes the earth to make one revolution on its axis. Where the day is too large a unit, the *hour* (1/24 of a day) is used. In turn, the hour is divided into *minutes* (1/60 of an hour), and *seconds* (1/60 of a minute). Where the day is too small, units such as the *week* (7 days), *year* (365 days), *decade* (10 years), *century* (100 years) and *millenium* (1000 years) are used.

Time is measured by clocks. In control and instrumentation systems, there are three basic types of clocks. A *spring-motor* clock measures time by doling out the energy of a coiled spring at intervals of fixed duration as determined by a balance wheel-and-escapement movement. The uncoiling of the spring causes the rotation of a set of gears which, in turn, causes the hands of the clock to move across a dial. Such clocks can measure time in intervals as small as about one-tenth of a second.

The spring-motor clock has been replaced by an *electric clock* in most control and instrumentation systems. A typical electrical clock consists of a small electric motor, running at a constant speed, that operates a set of

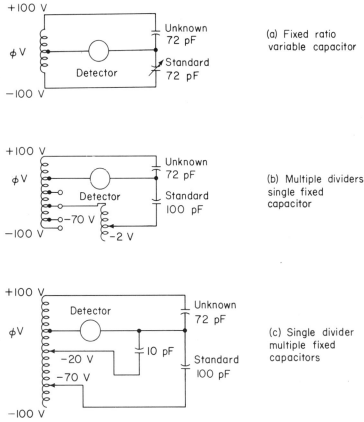

Figure 8-35 Methods of balancing capacitance in a transformer-ratio bridge.

gears to move the hands of the clock. The rotation of the motor, which is a *synchronous* type (Chapter 9), is in exact step, or synchronization, with the alternations of the a-c line from which it is operated. Since the power companies keep these alternations constant (at 60 Hz per second in most cases), the motor speed is constant. Electric clocks measure intervals as small as about one-hundredth of a second.

The most modern clock is an *electronic clock*. An electronic wrist watch is a familiar example of an electronic clock. Here, a stream of precisely timed pulses are generated by a quartz crystal oscillator. These pulses are counted up and down, as necessary, to count off seconds, minutes, and hours. In turn, the outputs of the seconds, minutes, and hour counters are displayed as electronic readouts. Operation of counters is discussed in Sec. 8-5 and 8-6. Here, we concentrate on *timers* and *timing functions* in control and instrumentation systems.

Timers are widely used in industry to initiate some process and then to stop it after a predetermined time interval. The duration of such intervals may be seconds, hours, or even days. For some applications, the time intervals are measured in microseconds or some other fraction of a second.

There are five basic types of timers in common use. An *interval timer* initiates a process immediately upon start of a timing period, keeps the process going during a preselected period, and stops the process when the period is complete. A *time delay timer* initiates a process at the end of a predetermined timed delay, and stops the process after another predetermined time lapse. A *recycling timer* starts and stops a process after a predetermined lapse of time, and immediately resets itself to repeat the cycle as long as the motor that drives it keeps running. A *time switch* opens and closes a switch at predetermined intervals during some time unit such as a day. (Timers used to control electric lights are a common form of time switch.) A *running time meter* records the total operating time for a piece of equipment.

8-4.1   Clock Timers

Clock-driven timers are generally used when the time intervals are about one second or longer, although there are exceptions. Figure 8-36 shows some typical clock timers. Such timers usually have a synchronous electric motor (Chapter 9) driving a train of gears. Often, the motor engages the gear train by means of a solenoid-operated clutch (Chapter 9). Although the motor is running continuously, the timing operation does not start until the motor and gear train are engaged. An electric pulse, produced by a switch or some other device, engages the solenoid of the clutch, engaging the motor and the gears, thus initiating the timing operation.

As the gears rotate, they carry with them a cam which, at the end of a predetermined time interval, engages an arm that opens or closes a switch; see Fig. 8-36(a). By positioning the cam in relation to the arm, the time lapse between the start and the point where the arm is engaged may be set. When the time interval is completed, the cam-operated switch deenergizes the solenoid of the clutch, thus stopping the timing operation. The cam automatically returns to its original position. In the recycling timer, the clutch is then immediately reenergized, and the cycle is repeated. In the running time meter of Fig. 8-36(b), the cam is geared to make one revolution for each time unit indicated by the meter. Then, as the cam makes one revolution, it closes a switch, sending a pulse to a register, thus advancing the register by one count.

8-4.2   Electronic Timer Basics

Where the time intervals involved are in the order of a few microseconds to a few minutes, electronic timers are generally used. Operation of most electronic timers is based on the *charge and discharge of a capacitor*. If a

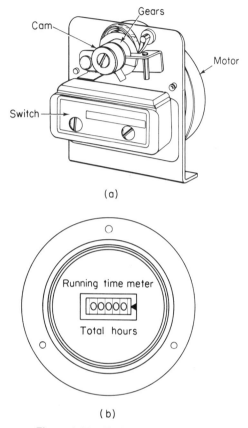

(a)

Running time meter

00000

Total hours

(b)

Figure 8-36 Typical clock timers.

capacitor is connected to a d-c voltage, the instant the circuit is completed a relatively heavy charging current flows. The current falls off quickly as the capacitor is being charged and stops completely when the capacitor is "fully charged." If a resistor is placed in series between a voltage source and a capacitor, as shown in Fig. 8-37, the resistor impedes the current flow. The time it takes the capacitor to reach "full charge" then depends on the values of the capacitor and the resistor. Theoretically, a capacitor never becomes fully charged, so it is accepted practice to consider a capacitor "fully charged" when it reaches 63% of the true full-charge value. The time that it takes to reach the 63% point is called the *time constant.*

Figure 8-37 shows the relationship of a time constant to capacitor charge. The reason that a capacitor never becomes fully charged, in theory, is that a counter charge (also known as a counter electromotive force, or cemf) is built up on the capacitor and opposes the current flow. The longer the time from the start of charging, the greater the counter force (resulting in an increasingly lower current flow). When current flow drops to 37% of

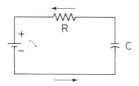

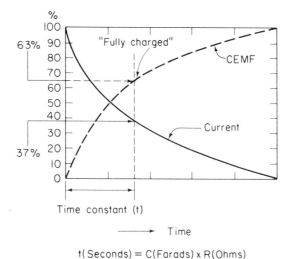

$$t(Seconds) = C(Farads) \times R(Ohms)$$

Figure 8-37 Relationship among current, CEMF, and time as a capacitor is charged.

the initial value, the counter force has reached 63% of its initial value and the capacitor is said to be fully charged for practical purposes.

The time constant (or time required to reach the 63% full-charge point or 37% current point) is calculated as follows:

$$t \text{ (in seconds)} = C \text{ (in farads)} \times R \text{ (in ohms)}$$

For example, let us determine the time constant of an *RC* circuit consisting of an 0.001 microfarad (0.000,000,001 farad) capacitor and a 1 megohm (1,000,000 ohms) resistor:

$$t = 0.000,000,001 \text{ farad} \times 1,000,000 \text{ ohms}$$
$$= 0.001 \text{ second} = 1,000 \text{ microseconds}$$

8-4.3 Electronic Time-Delay Relay

The principle of using capacitor charge and discharge through known resistance values can be readily adapted to control the timing function of an electronic timer, also known as an *electronic time-delay relay*. Such time-delay relays are particularly useful for industrial or other applications that require a time delay of definite duration (typically very short duration) after a switch is opened or closed. For example, in photographic printing, it is necessary to have a carefully controlled exposure time ranging from a fraction of a second to several minutes. An electronic time-delay must turn on the exposure or printing lights as a switch is operated, and automatically turn off the lights after a predetermined time interval. Older electronic time-delay relays used vacuum tubes. These have been largely superseded by solid-state time delay relays.

Figure 8-38 shows a basic electronic time-delay relay circuit that uses a capacitor charge to control a time interval. Operation of the circuit is started when push button switch *S* is closed (pressed), which shorts out capacitor *C*, effectively removing *C* from the circuit. Closing switch *S* also places a negative voltage (from the battery) on the base of transistor *Q*, and forward biases the base-emitter junction. (As discussed in Sec. 7-3.3, a PNP transistor is "turned on" when the base is more negative than the emitter. This is known as "forward biasing" a transistor or "turning on" a transistor.) With transistor *Q* turned on, collector current flows through the coil of relay *K*, closing the relay contacts. In turn, these contacts control operation of the circuit being timed by the time-delay relay. For example, closure of the relay contacts could turn on photographic exposure lights.

When push button switch *S* is released, it opens, and capacitor *C* starts to charge up, with the polarity indicated in Fig. 8-38 (the base side of *C* starting to go positive). As *C* charges, its counter force opposes the base-emitter forward bias produced by the battery. When *C* is charged sufficiently, the counter force overcomes the forward bias on *Q*, and *Q* stops conducting. This removes current from relay *K*, and the relay contacts open, removing power to the circuit being controlled or timed (such as turning off photographic exposure lights).

The time required for *C* to charge up to the point where *Q* becomes nonconductive after *S* is opened depends on the capacitance of *C*, the resistance

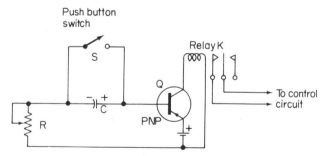

Figure 8-38 Basic electronic time-delay relay (for long time intervals).

of *R,* and the base-emitter resistance of *Q.* Since the capacitance of *C* and the base-emitter resistance of *Q* are fixed, the time constant of the circuit may be controlled by adjustment of *R.*

The circuit of Fig. 8-38 is particularly effective for time delays longer than the time required to press and release a push button (typically 1 second or longer). When the time interval must be a fraction of 1 second, a circuit similar to that shown in Fig. 8-39 is more effective. The Fig. 8-39 circuit uses capacitor discharge to control the time interval. Normally, push button switch *S* is in position *A.* This charges capacitor *C* up to the full battery voltage. Transistor *Q* is not turned on, since there is no voltage applied to the base and no current flow through the coil of relay *K* (with the possible exception of some leakage current from emitter to collector). In any event, the current is not sufficient to operate relay *K,* and the contacts of relay *K* remain open.

Operation of the Fig. 8-39 circuit is started when push button switch *S* is moved to position *B* (pressed and held). The positive voltage of *C* places a forward bias on the base-emitter junction of *Q* (which is an NPN transistor in this case). With transistor *Q* turned on, collector current flows through the coil of relay *K,* closing the relay contacts. In turn, these contacts control operation of the circuit being timed (turn on of exposure lights, etc.). Since *C* is disconnected from the battery, *C* starts to discharge through two parallel paths: through *R*1, and through *R*2 and the base-emitter junction. As *C* discharges, the forward bias on *Q* decreases, as does the collector current through the coil of *K.* When this discharge of *C* is sufficient, the collector current falls below the value required to keep *K* energized, and the relay contacts open, removing power to the circuit being timed (turn off of exposure lights, etc.).

The delay interval is the time lapse between pressing switch *S* to position *B* (thus closing the relay contacts), and the deenergizing of the relay (thus

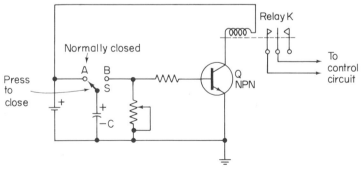

Figure 8-39   Basic electronic time-delay relay (for very short time intervals).

opening the relay contacts). This time interval is determined by the capacitance of C, the resistances of $R1$ and $R2$, and the base-emitter resistance of Q. As before, the time interval of the circuit can be controlled by adjustment of $R1$. Since the time interval is typically a fraction of 1 second, push button switch S need be pressed and held for about 1 second.

8-5 COUNTERS

Counting is a form of measurement. There are two types of counting found in control and instrumentation systems. One type of counting is used to obtain an *indication of quantity* such as a number of units or of actions performed. An example is counting the number of units passing by a given point on a conveyor belt. The other kind of counting is used as a step in control systems where a certain *action is initiated or terminated* after a certain number of units or actions has been counted. An example is the stopping of a conveyor belt after a certain number of units have passed a given point. No matter what the application, there are three basic types of counters used in control and instrumentation systems. These include mechanical and electrical counters, which are discussed in the following paragraphs, and electronic counters, which are covered in Sec. 8-6.

8-5.1 Mechanical Counters

The most common form of mechanical counter is shown in Fig. 8-40. Such counters are also known as *registers*. A typical mechanical counter consists of a set of wheels, each coupled to the next set at the left through a 10-to-1 reducing gear. The rim of each wheel is divided into 10 equal parts, and these parts are numbered consecutively from 0 to 9. The numbers appear at a window on the register.

In the counter or register of Fig. 8-40, the extreme right-hand wheel advances one-tenth of a revolution each time the stroke arm is actuated. That is, the right-hand wheel advances one number. It takes 10 strokes to make one complete revolution of the right-hand wheel (called the *units* wheel). Thus, the number on the units wheel indicates the number of strokes (from 0 to 9) that have occurred for a given series of actions (such as objects passing by a given point on a conveyer, with each object actuating the stroke arm once).

Each time the units wheel makes one revolution and returns to the 0 position, the *tens* wheel (at the left of the units wheel) advances one-tenth of a revolution. Thus, each number on the tens wheel represents 10 strokes. Similarly, the next wheel to the left is the *hundreds* wheel, and each of its numbers represents 100 strokes. The next wheel to the left is the *thousands* wheel, etc. By observing the numbers in the register window, the number of strokes can be determined at a glance. A register with six wheels (which is usually maximum for mechanical counters) can count up to 999,999. When

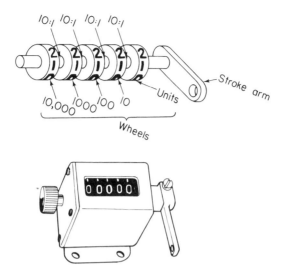

Figure 8-40 Basic mechanical counter or register.

the limit of the register is reached, the next stroke returns the count to 000,000, and the process is repeated. Most mechanical counters have a *reset* function (generally a knob or button) that permits the count to be reset to zero on all wheels simultaneously, without having to go through each step.

Another type of mechanical counter is the *revolution counter* or *register*. This type of counter is similar to the stroke counter shown in Fig. 8-40, except that the wheels are actuated by the revolution of a shaft rather than by actuation of a stroke arm. The linear-footage counter used by estimators is a classic example of the revolution counter. Here, the counter is operated by a measuring wheel attached to the drive shaft. After being reset to zero, the counter is rolled along over the area to be measured. This causes the wheel to turn the shaft which, in turn, operates the counter wheels and indicates the total linear footage. In control applications, the revolution counter is generally stationary, with the shaft being driven by some mechanism that is part of the control operation (such as a conveyor belt, motor, etc.).

One of the problems of a mechanical counter is that it must be located near the work area (or where the units or actions are being counted). It is not always convenient or desirable to read the counter at that location. There are circuits that can transmit a count or reading to a remote location. However, if such circuits are required, the mechanical counter is often replaced by an electrical counter.

8-5.2 Electrical Counter

An electrical counter or register, which resembles a mechanical register except that the units wheel is activated by an electromagnet, is able to *count electrical pulses.* Each pulse of current flowing through the coil of the electromagnet advances the count by one unit. The pulses can come from any source and can be transmitted from a work area to a remote location or possibly a control station. The pulses can be generated by a mechanical switch or some similar actuator. When the objects to be counted are fragile or must not be touched for some reason, a photocell described in Chapter 5 can be used to produce the pulses.

Figure 8-41 shows a typical electrical counter and how it can be used in conjunction with a photocell detector to count objects passing by on a conveyor belt. As discussed in Chapter 5, a photocell detector can be arranged to operate a relay when light is applied or when light is removed. Closing of the relay contacts provides pulses to the electrical counter. As shown in Fig. 8-41, a photocell is mounted on one side of a moving conveyor belt and a fixed light source is mounted on the other side of the belt. A light beam is interrupted as objects on the belt pass between the photocell and light source. As a result, an electrical pulse is sent from the photocell relay to the electrical counter, advancing the count by one.

8-6 ELECTRONIC COUNTERS

Mechanical and electrical counters are limited to about 15,000 counts per minute. Beyond that, an electronic counter such as shown in Fig. 8-42 is required. As in the case of digital meters, it is necessary to have a knowledge

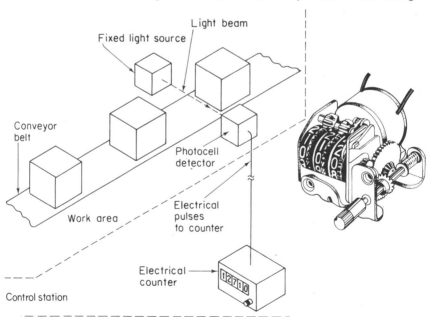

Figure 8-41 Typical electrical counter application.

of digital logic circuits, particularly decades, decoders, and numerical readouts, to fully understand electronic counters. Operation of these circuits is presented in block form here, and the circuits are discussed (relative to counters) as necessary to understand operation of an electronic counter. Full descriptions of these devices contained in the author's *Logic Designer's Manual* referenced in Sec. 7-3.5.

8-6.1 Basic Electronic Counter

An electronic counter operates by comparing an unknown frequency or time interval to a known frequency or known time interval. The counter presents the information in an easy-to-read, unambiguous numerical display or readout. The readout of most electronic counters is similar to that of a digital meter (Sec. 8-2). In fact, a digital meter is essentially an electronic counter, plus a conversion circuit for converting voltage to a series of pulses.

The accuracy of an electronic counter depends primarily on stability of a known frequency, which is usually derived from an internal oscillator. The accuracy of the oscillator can be checked against broadcast standards. Thus, an electronic counter can become a frequency or time standard. Counters are often used with accessories and in systems. For example, there are electronic counters that retain their counts for automatic recording of measurements, digital clocks that control measurement intervals and supply time information for simultaneous recording, digital-to-analog converters for high-resolution analog records of digital measurements, and scanners that can receive the outputs from several electronic counters for entry into a single recording device. Also, there are magnetic and optical tachometers for revolutions per second measurement (such as described in Chapter 2) that are designed to provide inputs to frequency counters.

The electronic counter shown in Fig. 8-42 is used mostly for the

Figure 8-42   Heathkit IM-4100 frequency counter. (Courtesy of Heath Co.)

measurement of *frequency* but can also be used to *totalize* counts or to measure *time intervals* and *periods*. Note the simplicity of its controls. Once power is turned on, an operator has only to select the function, the level of the input signal to be measured, and the time base or rate at which the counting operation is to be performed. Operation of these controls and the functions performed by the counter are discussed in the following paragraphs.

8-6.2 Counter Circuit Elements

Although there are many types of electronic counters, all counters have several functional sections in common. These sections are interconnected in a variety of ways to perform the different counter functions.

All electronic counters have some form of *counter and readout assemblies.* Early instruments used *binary counters* and *readout tubes* that converted the binary count to a decade readout. Such instruments have (generally) been replaced by *decade counters* that convert the count to binary-coded-decimal (BCD) form, *decoders* that convert the BCD data to decade form, and readouts (often solid state) that display the decade information directly. (BCD, decade counters, and decoders are discussed in Sec. 8-6.9. Readouts are discussed in Chapter 10.) For the purpose of discussion here, one readout is provided for each digit. Thus, eight readouts provide for a count (or time interval) up to 99,999,999.

All electronic counters have some form of *main gate* that controls the count start and stop with respect to time. Usually, the main gate is some form of AND gate, *where two or more simultaneous inputs are required to produce an output.*

All electronic counters have some form of *time base* that supplies the precise increment of time to control the gate for a frequency or pulse train measurement. Usually, the time base is a *crystal-controlled oscillator.* The accuracy of the counter is dependent upon the accuracy of the time base, plus or minus one count. For example, if the time base accuracy is 0.005%, the overall accuracy of the counter will be 0.005%, plus or minus one count. The one-count error arises because the count may start and stop in the middle of an input pulse, thus omitting one pulse from the total count. Or part of a pulse may pass through the gate before it is closed, thus adding a pulse to the count.

Most electronic counters have *dividers* that permit variation of gate time. That is, the dividers convert a fixed-frequency time base to several other frequencies.

In addition to these four basic sections, most electronic counters have attenuator networks, and amplifier and trigger circuits to shape a variety of input signals to a common form, as well as logic circuits to control operation of the instrument. The various modes of electronic counter operation are described in the following sections.

8-6.3 Totalizing Operation

Electronic counters can be operated in a totalizing mode with the main gate controlled by a manual START/STOP switch as shown in Fig. 8-43. With the switch in START position (gate open), the counter assemblies totalize input pulses until the main gate is closed by the switch being changed to STOP. The counter display then indicates the number of input pulses received during the interval between manual START and manual STOP. Generally, totalizing can be remotely controlled. For example, the input pulses can come from a photocell detector counting objects passing by on a conveyor belt (Sec. 8-5.2). The manual switch and readouts are located at a remote control station. The operator can then start and stop the count at will by operating a START/STOP switch. For some control systems, it is necessary to know how many objects are passing during a given period of time (say for 1 hour). In other instances, it is more important to maintain the counting operation (switch in START position) until a given number of objects pass, without regard to time.

In a totalizing operation, the main gate is a form of AND gate. An AND gate requires two (or more) like inputs to produce an output. For example, if the AND gate operates with positive inputs (known as a "positive true AND gate"), two positive inputs produce an output. When a manual control switch is set at START, a positive voltage is applied to one input of the AND gate. The input pulses (also positive) are applied to the other AND gate input. Thus, there is one output pulse to the counter/readout for each input pulse applied. When a manual control switch is set at STOP, the positive voltage is removed from the AND gate input. Further input pulses provide only one input to the AND gate, resulting in no output pulses to the

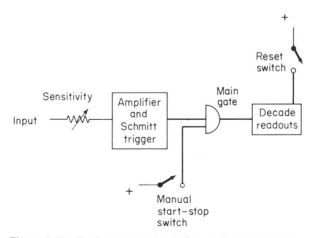

Figure 8-43 Basic counter circuit for totalizing operation.

201

counter/readout. In some counters, the counter/readouts have the ability to totalize in either a positive or negative direction.

As shown in Fig. 8-43, input signals or pulses to be counted are applied to the AND gate through an *attenuator* network, an *amplifier,* and a *Schmitt trigger.* The attenuator network (sometimes known as a *sensitivity* network or switch) determines the level of input signals that will produce a count. (All input signals above a certain level are counted; all signals below that level are not counted for a given setting of the attenuator switch.) The amplifier (not found in all counters) raises the level of input signals to a point where they can be counted. The Schmitt trigger produces an output (usually a square wave) of fixed amplitude and duration for each input signal (above a given level). The output pulse remains fixed, regardless of the input signal shape. Thus, the Schmitt trigger is useful in counter circuits where the input signals to be counted are of poor quality.

Reset function.   Counter/readout assemblies must be reset. If not, a new count is added to a previous count each time the gate is opened. Manual reset is accomplished by pushing a RESET button, which applies a voltage to all of the counters simultaneously, setting them at a position that produces a numerical zero on each readout. For other modes of operation (such as frequency measurement, time interval, or period measurement), reset is accomplished by a pulse applied to all counters at regular intervals. This pulse is developed by a low-frequency oscillator (usually 2 or 3 Hz). Therefore, two or three counts or "samples" are taken each second. The pulse oscillator (often called the *sample rate* oscillator) is adjustable in frequency on some counters.

8-6.4   Frequency Measurement Operation

In the frequency measurement mode, the signals to be counted are first converted to uniform pulses by a Schmitt trigger as shown in Fig. 8-44. The pulses are then routed through the main gate and into the counter/readout, where the pulses are totalized. The number of pulses totalized during the "gate-open" interval is a measure of the *average input frequency* for that interval. For example, assume that a gate is held open for 1 second and the count is 333. This indicates a frequency of 333 Hz.

The count obtained, with the correct decimal point, is then displayed and retained until a new sample is ready to be shown. The TIME BASE selector switch selects the gating interval, thus positioning the decimal point and selecting the appropriate measurement units. As shown in Fig. 8-44, the TIME BASE selector selects one of the frequencies from the time base oscillator. If the 1 kHz signal (directly from the time base) is selected, the time interval (gate-open to gate-close) is 1 mS (millisecond). If the 100 Hz signal (from the first decade divider) is chosen, the measurement time inter-

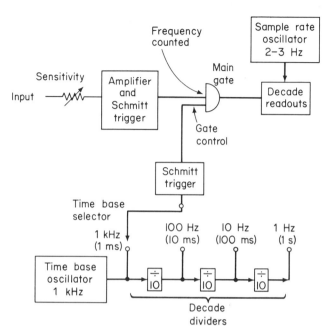

Figure 8-44 Basic counter circuit for frequency measurement operation.

val is 10 mS. The 10 Hz signal (from the second decade divider) produces a 100 mS interval. The 1 Hz signal from the third decade divider produces a 1 second interval.

8-6.5 Period Measurement Operation

Period is the inverse of frequency (period = 1/frequency). Thus, period mode measurements are made with the input and time base connections reversed from those for frequency measurement as shown in Fig. 8-45. The unknown input signals to be counted control the main gate time, and the time base frequency is counted and read out. For example, if the time base frequency is 1 MHz, the indicated count is in microseconds: a count of 80 indicates that the gate has been held open for 80 μs (microseconds).

The accuracy and resolution of period measurement can be increased by *period averaging* as shown in Fig. 8-46. The connections are the same as for regular period measurement, except that the signals to be counted are lowered in frequency by dividers, thus extending the gate-open period. For example, if the input signal is 1 kHz, the period is 1 mS with a regular period measurement. If the time base is 1 MHz, the count is 00001000 on an 8-digit readout. If the period average method is used and the input frequency is reduced to 1 Hz as shown in Fig. 8-46, the period is 1 second and the

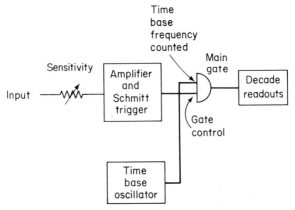

Figure 8-45 Basic counter circuit for period measurement operation.

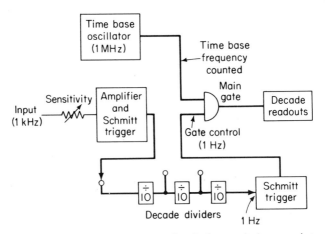

Figure 8-46 Basic counter circuit for period averaging measurement operation.

count is 01000000 on the same 8-digit readout. Thus, the resolution is increased by 1000.

8-6.6   Time-Interval Measurement

Time-interval measurements are essentially the same as period measurements. However, a time-interval mode concerns time between two events, rather than the repetition rate of signals to be counted. Counters vary greatly in their time-interval measuring capability. Some counters measure only the duration of an electrical event; others measure the interval between the start of two pulses.

A basic time-interval measurement circuit is shown in Fig. 8-47. Note

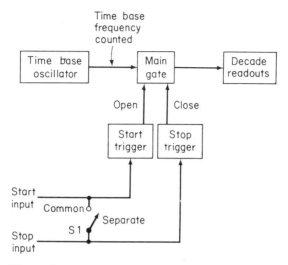

Figure 8-47 Basic counter circuit for time-interval measurement operation.

that the time-base signals are counted and read out when the gate is open. Control of the gate is accomplished by two trigger circuits that receive their inputs from the signals being measured. With switch S in the SEPARATE position, the two triggers receive inputs from separate lines. With switch S in the COMMON position, the two triggers receive inputs from the separate lines. Either way, the opening and closing of the gate permits a measurement of the time interval between the two triggers which, in turn, are produced by the two events being timed.

8-6.7 Clock Operation

Although all electronic counters do not have a clock function, the same basic circuits are used for an electronic clock. A typical electronic clock circuit is shown in Fig. 8-48. (This circuit is similar to that used for digital watches and clocks.) As shown, the circuit consists essentially of a highly stable crystal oscillator time base (the frequency of which is divided down to produce 1 pulse per second pulses) and a digital readout similar to that of an electronic counter.

Once set, the clock operates continuously with the count increasing by 1 second for each pulse. The clock is set by introducing a series of fast pulses (usually from the dividers at some rate much faster than 1 pulse per second) into the counter until the correct time is indicated. Then the counter input is returned to 1 pulse per second. In addition to a visual readout, some clocks used in control systems provide an electrical output that can be applied to other equipment.

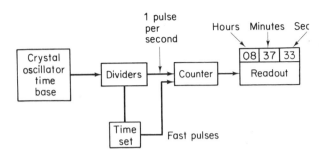

Figure 8-48 Basic circuit for electronic clock.

8-6.8 Counter/Readout and Divider Circuit Operation

Most modern counters use decade counters (or *decades*) comprised of four binary counters that convert a count to a BCD code, decoders for conversion of a code to decimal form, and readouts that display information directly in decimal form. The same circuit used for decade counters (four binary counters) is also used as *dividers*. Therefore, it is necessary to understand operation of the basic decade before going into how the decade is used in electronic counters or clocks.

Decades. Decade counters (or decades) serve two purposes in electronic counters. First, the decade will divide frequencies by 10. That is, the decade produces one output for each 10 input pulses or signals, thus permitting several frequencies to be obtained from one basic frequency. For example, a 1 MHz time base can be divided to 100 kHz by one decade divider, to 10 kHz by two decade dividers, to 1 kHz by three decade dividers, etc. When decades are used for division they are sometimes called *scalers,* although *dividers* is a better term. The second purpose of a decade is to convert a count to a BCD code. The division function of a decade is discussed here first.

The basic unit of a decade divider is a 2:1 scaler, called a *binary counter.* This unit uses a *bistable multivibrator* or a logic *flip-flop* (FF), depending on design. A basic flip-flop is shown in Fig. 8-49. Although the circuit shown uses discrete components, decade flip-flops are most often found in IC form, with all four FF's in one package. Sometimes several decades are found in one package.

The flip-flop of Fig. 8-49 is made up of two cross-coupled AND gates. Such a FF has two stable states, with the gate A output positive and the gate B output negative and vice versa. The first input pulse flips the circuit from one state to the other. The second input pulse flops the circuit back to its original state; hence the name *flip-flop.* Each time the circuit is flipped from

206

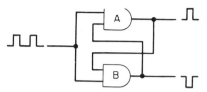

Figure 8-49 Basic flip-flop (binary counter on 2:1 scaler).

one state to the other and back again (requiring two input pulses) a single (complete) output pulse is produced. The output from the FF may be taken from either gate *A* or gate *B*. The important point to remember is that the outputs will always be in opposite states, and that it takes two input pulses to produce one output pulse at either output.

The output pulses of one FF may be applied to the input of another similar FF for further frequency division. This condition is called *cascading*. A basic binary counter has a cascaded chain of four FF's as shown in Fig. 8-50. The count of this chain is 16. That is, for every 16 input pulses to be counted, the output of FF1 is 8, of FF2 is 4, of FF3 is 2, and of FF4 is 1. When the count is to be divided by 10, some of the FF outputs are *fed back* to cancel the pulses, as shown in Fig. 8-51. Here, for every 10 input pulses to be counted, the output of FF1 is 5, of FF2 is 3, of FF3 is 2, and of FF4 is 1.

Decade-to-BCD conversion. The decades shown in Figs. 8-49 through 8-51 are used for division. The same basic circuit can be used to convert a series of pulses into a binary code such as BCD. To understand this function, it is necessary to understand the binary counting system and how the BCD code relates to that system. Thus, we now review these subjects before discussing decade-to-BCD conversion.

In the binary system, all numbers are combinations of only ones and zeros (1 and 0) rather than zero through nine, as in the familiar decimal system. Consequently, instead of requiring 10 different values to represent one digit, circuits using the binary method need only two values for each digit. In electronic counters, and other digital electronic equipment, these values are easily indicated by the presence or absence of a signal (or pulse) or by positive and negative signals or even by two different voltage levels. Thus, binary is adaptable to any electronic device.

In binary, the value of each digit is based on 2 and the powers of 2. In a binary number, the extreme right-hand digit is multiplied by 1, the second-from-the-right digit is multiplied by 2, the third-from-the-right digit is multiplied by 4, and so on. This can be displayed as follows:

2^8	2^7	2^6	2^5	2^4	2^3	2^2	2^1	2^0
256	128	64	32	16	8	4	2	1

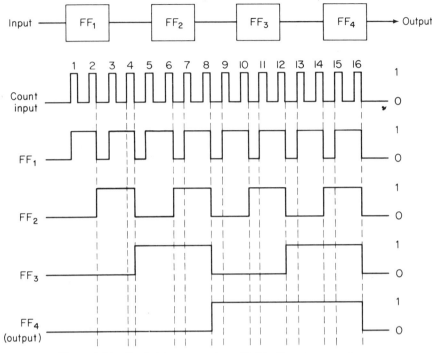

Figure 8-50 Basic binary counter with flip-flops in cascade.

In binary, if the digit is zero, its value is zero. If the digit is one (1), its value is determined by its position from the right. For example to represent the number 77 in binary form, the following combination of zeros and ones is used:

256	128	64	32	16	8	4	2	1
0	0	1	0	0	1	1	0	1

$$64 + 0 + 0 + 8 + 4 + 0 + 1 = 77$$

which means that 1001101 in pure binary form equals 77.

The binary coded decimal (BCD) system combines the advantages of the binary system (the need in digital electronic circuits for only two states, one and zero) and the convenience of the familiar decimal representation. In the BCD system, a number is expressed in normal decimal coding, but each digit in the number is expressed in binary form.

For example, the number 37 in BCD form appears as:

	TENS DIGIT	UNITS DIGIT
Decimal:	3	7
BCD:	0011	0111

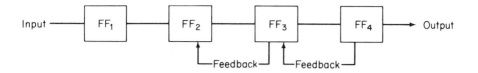

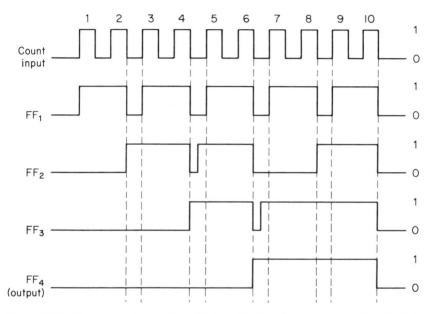

Figure 8-51 Binary decade counter with four flip-flops in cascade, and feedback from FF3 to FF4 to FF3.

Note that four bits of information are needed for each digit. In general, four bits yield 16 possible combinations. However, in BCD only 10 combinations are needed.

Figure 8-52 shows a decade circuit capable of converting a series of pulses into the BCD code. One FF is used for each of the digits. Input pulses are fed to the 1-FF. The 2-, 4-, and 8-FF's are cascaded and receive pulses after the 1-FF.

At the beginning of a count, the 8421 lines (representing the 8-, 4-, 2-, and 1-FF outputs, respectively) are at negative, which is represented by a 0. In a typical counter, the decades are set (or reset) to this condition by the application of a voltage or pulse produced by a reset button, or by the sample rate generator, or by pulses from the decades at the end of a 10 count.

When the first pulse in the count is applied, the 1-FF changes state. The 1-line becomes positive, represented by a 1. When the second pulse is applied, the 1-FF again changes state. The 1-line goes negative (0), and the 2-line goes positive (1).

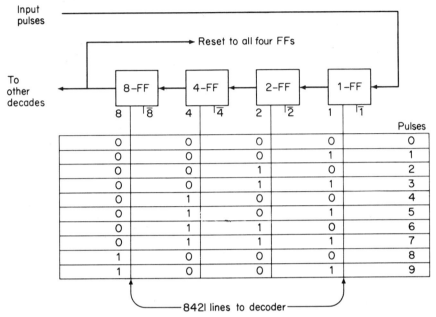

Figure 8-52 Binary decade circuit for converting a series of pulses into an 8421 binary code.

With the third pulse applied, the 1-FF goes positive (1), but the 2-FF remains positive. Remember that the 2-FF changes state for each complete cycle (or two-state change) of the 1-FF.

When the fourth pulse is applied, the 1-FF goes to 0, as does the 2-FF. This causes the 4-FF to change states (the 4-line goes to 1).

This process is repeated until a nine count is reached. At that point, the 8-FF moves from 1 to 0. This 0 output is returned to the reset line and serves to reset all of the FF's to the negative or 0 state. The output from the 8-FF can also be applied to the 1-FF of another decade. Any number of decades can be so connected. One decade is required for each readout numeral (in most cases).

BCD-to-7-segment readout conversion. Most electronic readouts now use some form of 7-segment display as shown in Fig. 8-53. The numerals formed on the display of Fig. 8-42 use the 7-segment system. The numerals (from 0 through 9) are formed when the corresponding segments are turned on. For example, to form the numeral 8, all segments are turned on simultaneously. To form the numeral 3, all segments *except e and f* are turned on. The method of turning on the segments depends on the type of display used. The most common forms of numerical readouts are liquid crystal displays (LCD), light-emitting diodes (LED), gas-discharge, fluorescent,

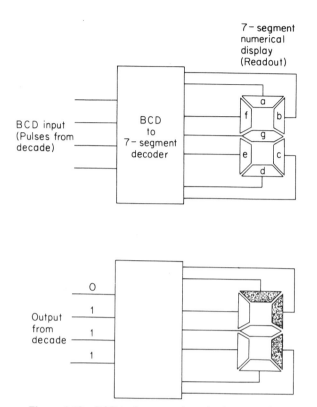

Figure 8-53 BCD-to-7 segment readout conversion.

and incandescent. These displays are discussed in Chapter 10. Here, we are concerned as to how the segments are selected to form the correct numeral.

Generally, the individual segments are turned on when a voltage is applied to a selected segment. In the readout of Fig. 8-53, the voltage is applied by a BCD-to-7-segment decoder in response to pulses from a decade such as shown in Fig. 8-52. For example, when seven pulses to be counted are applied to the decade in Fig. 8-52, the output of the decade to the decoder is 0111, which is 7 in binary or BCD. The decoder of Fig. 8-53 then converts this to voltages on segments *a, b,* and *c,* with all other segments receiving no voltage (turned off). Segments *a, b,* and *c* are turned on, and the numeral 7 is formed.

Operation of the decoder is not discussed here for two reasons. First, the discussion requires a knowledge of digital electronics. Second, present-day decoders are in IC form, where access to the internal circuit is not available (even if you understood its operation).

9

BASIC CONTROL DEVICES

As discussed in Chapter 1, the final stage of most control systems includes (1) a switch which may be opened or closed, (2) a valve which may be opened or closed or adjusted to some position between those two extremes, (3) an electromagnetic device which may be energized by an electric current to perform some mechanical or electrical function, or (4) a motor which may be started, stopped, or reversed, or whose speed may be varied while running.

Between the transducer (which functions as the primary sensing element and initiates operation of a control system) and the final control element there may be several of a host of devices, each performing a definite function in the system. Such devices are switches, relays, solenoids, motors, valves, actuators, electron tubes, semiconductor (solid-state) devices, and many more, in various combinations. In this section, we examine some of the commonly used control devices, explain their operation, and see how they fit into a control system.

9-1 SWITCHES AND RELAYS

Switches and relays are referred to as *contactors,* since their primary function is to complete an electrical circuit, which is done by opening and closing contacts, manually in the case of switches, electrically in the case of relays. Switches and relays are classified according to the action they perform. Each movable contact is called a *pole*. When only one movable contact is actuated, the switch or relay is *single-pole*. A *double-pole* switch or relay has two movable contacts that are actuated simultaneously. The positions in which the movable contacts may be placed to touch the fixed contacts are called *throws*.

Figure 9-1 shows the physical construction and electrical symbol for three basic switches: *single-pole, single-throw* (SPST), *double-pole, single-throw* (DPST), and *double-pole, double-throw* (DPDT). Section 9-1.1 describes some typical switches found in present-day control systems.

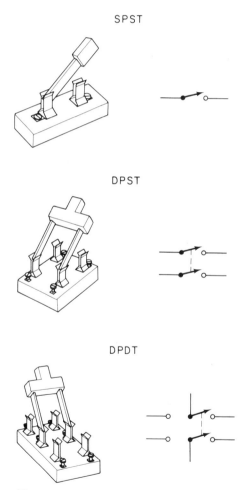

SPST

DPST

DPDT

Figure 9-1 Basic SPST, DPST and DPDT switches.

9-1.1 Control System Switches

Figure 9-2 shows one of the most common types of industrial control switch. It is a DPST *knife* switch enclosed in a metal box and actuated by an external handle. The poles or contacts have provisions for fuses. Such switches consist of hinged metal blades (the movable contact) and metal clips (the fixed contacts) into which the blades fit. This kind of switch is often used as a main power switch.

Figure 9-3 shows a DPDT *slide* switch, which is actuated by sliding a button from side to side, and two versions of *toggle* switches.

Figure 9-4 shows an SPST *mercury* switch, which is closed (contact completed) when liquid mercury shorts a set of contacts. The mercury and contacts are contained in a glass or plastic tube, and the switch is thrown from one position to the other by tilting the tube. The tilting action, which causes the mercury to move through the tube and close or open the contacts, may be performed manually by means of a toggle mechanism. Mercury switches

Figure 9-2 Fusable industrial control switch. (Courtesy of Square D Co.)

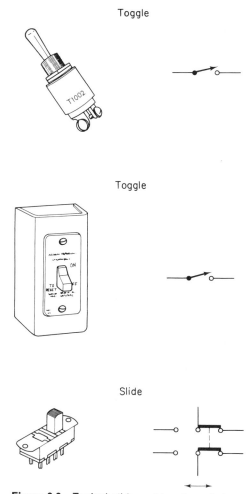

Toggle

Toggle

Slide

Figure 9-3 Typical slide and toggle switches.

can also be operated automatically by clock mechanisms or any other control device capable of producing motion. It is possible to form an SPDT switch by using two mercury switches as shown in Fig. 9-4. When the mercury tubes are tilted to the right, the top set of contacts are closed, and the bottom set of contacts are opened.

The use of mercury switches is widespread in industrial control systems for several reasons. Since the contacts are enclosed, they do not produce sparks (minimizing the danger of an explosion), do not corrode (since they are not exposed), and are protected from dust and dirt. Mercury switches can also be made very small and light, have no wearing parts, and are noiseless. However, because they are glass-enclosed, they are more delicate than all-metal switches.

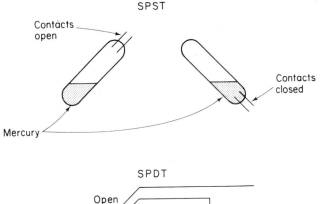

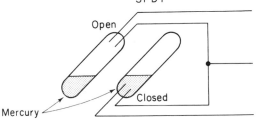

Figure 9-4 SPST and SPDT mercury switches.

Figure 9-5 shows a typical *push-button* switch. Generally, push-button switches are used if a circuit must be opened or closed *momentarily* and are often used to control relays. As shown by the symbols in Fig. 9-5, if the contacts are *normally-open* (NO), pressing the button closes the contacts. Pressing the button of a *normally-closed* (NC) switch opens the contacts. In either case, when pressure on the button is released, the switch springs back to its original position. As shown, one button can control more than one set of contacts, and one set of contacts can be opened when another set is closed.

Figure 9-6 shows the *snap-acting* switches, a *lever type* and a *plunger type*. These switches are similar to push-button switches in action (the switch contacts spring from one position to another) but are not manually operated. Generally, the switches are used as *limit* switches to limit the movement of machinery and other moving devices. When moving devices reach desired limits, they press against a switch plunger or lever, closing or opening a circuit to some device (such as a motor, solenoid, or relay), which stops the movement. Snap-acting switches are also used as actuators for electrical counters as discussed in Sec. 8-5.2.

Figure 9-7 shows a *rotary* switch, where a movable contact is in the form of a rotating arm which successively makes contact with a set of fixed contacts arranged in a circle. The arm and fixed contacts generally are located on an insulated wafer called a *deck*. The decks may be *ganged* so that several arms may be rotated simultaneously over their set of fixed contacts by means of a single control rod. Heavy-duty rotary switches, called *drum*

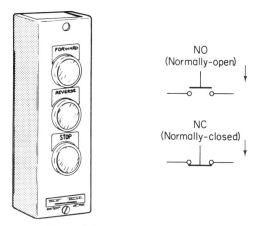

Figure 9-5 Push-button switch.

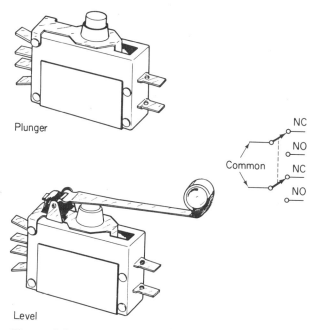

Figure 9-6 Plunger-type and lever-type snap-acting switches.

switches, are often used in industrial control systems for operation of electrical motors.

Any of the switches described thus far can be used in special applications. For example, a snap-acting switch can be actuated by a pressure sensor (bellows) or temperature sensor (bimetallic strips).

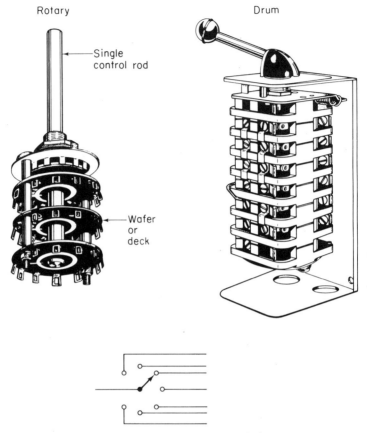

Figure 9-7 Rotary and drum switches.

9-1.2 Control System Relays

Figure 9-8 shows the physical construction and symbols for some typical *electromagnetic relays*. Such relays are essentially a set of contacts operated by an electromagnet (a coil of wire around a soft-iron core). The contacts can be normally-open or normally-closed and are operated when electrical current is applied through the coil. The current is supplied by a *control circuit*, usually involving a switch. When the contacts are opened or closed, they supply current to or operate a *controlled circuit*. When normally-open contacts close, they are said to *make*, and when normally-closed contacts open, they *break*.

As current from the control circuit flows through the turns of the coil, the relay is *energized*, and the resulting magnetic field around the core attracts the armature (which operates the contacts). As a result, the contacts

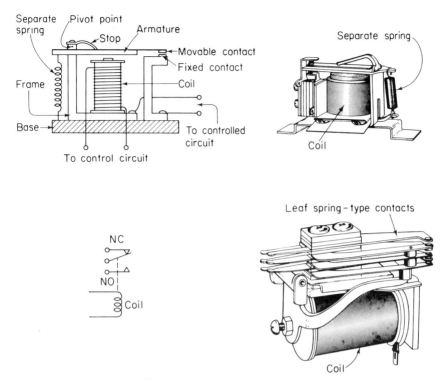

Figure 9-8 Typical electromagnetic relays.

break or make, and the controlled circuit is operated. When the flow of current through the coil stops, the relay is *deenergized,* and the spring returns the contacts to their normal position. Spring action can be accomplished by spring-type contacts in some relays. A separate spring is used in other relays. As shown in Fig. 9-9, the contacts can be arranged to make, break, break and make, make-before-break, and break-before-make, break.

Figure 9-10 shows a basic relay control system. Here, the control circuit consists of the power source, control switch, and relay coil. The controlled circuit consists of another power source, the load (or device being controlled), and the relay contacts. Figure 9-10 shows the two major advantages of a relay: *remote operation* and *control of large currents with small currents.* For example, assume that the load is an electrical device (say a bank of high-wattage lights) that must be operated from a remote location. If the lights are connected directly to a switch, a very heavy-duty switch must be used, and the wiring between the switch and lights must also be heavy duty. This results in a considerable power loss in the wiring and the possibility of early switch contact burnout. With the circuit of Fig. 9-10, the relay can be located as near as possible to the lights, and the switch can be located at any

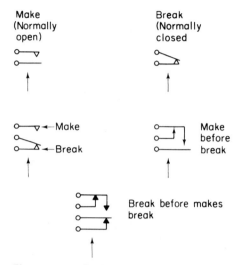

Figure 9-9 Typical relay contact arrangements.

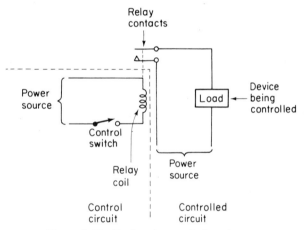

Figure 9-10 Basic relay control system.

convenient remote location. Thus, the switch and control circuit wiring can be of the low-current type, since the only current they must handle is that drawn by the relay coil (typically a few milliamperes or less).

Sparking at the contacts is a problem when relays are used in heavy-duty control sytems. As the contacts start to make, a surge of heavy-duty current causes a spark that may oxidize or pit the contact points. As the contact points break, the voltage of the controlled circuit causes another spark. One way to minimize this problem is to add a spark-reducing network (series resistor and capacitor) across the contacts as shown in Fig. 9-11. Resistor *R*

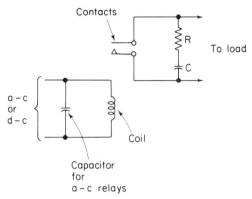

Contacts

R

To load

C

a – c
or
d – c

Coil

Capacitor
for
a – c relays

Figure 9-11 Circuits to minimize contact sparking and relay chatter.

reduces the make current surge. Capacitor *C* reduces the break voltage, since a certain amount of voltage is required to charge the capacitor when the contacts open. Another way to minimize the effects of contact sparking is to use metals that are resistant to oxidation and have high melting points for the relay contact points. Silver, tungsten, and paladium are typical metals used for high-quality relay contacts.

Also, as shown in Fig. 9-11, relays can be operated with either direct or alternating current. There are no particular problems with d-c relays, but when an a-c power source must be used for the control circuit, the relay may be subject to *relay chatter,* which is caused by the rapid alternations of the magnetic field, resulting in armature and contact vibration. Special construction is used for a-c relays, such as laminated frames, to reduce vibration. Also, as shown in Fig. 9-11, a capacitor can be placed across the coil of a-c relays. When the coil is energized, the capacitor is charged simultaneously. In between half-cycles, as the current goes from one alternation to the next, the capacitor discharges through the coil, thus keeping the coil energized and preventing chatter. Another more obvious alternative is to use a semiconductor diode (Sec. 7-3.2) to rectify the a-c power source to a d-c source for the control circuit.

Latching relay. The relay applications described thus far are *ON-OFF functions.* In many cases, a relay must remain energized after a control circuit is turned off. Using the previous example of a relay that controls lights at some remote location, the control switch must be held closed to keep the lights on. Obviously, this is not practical in many applications. Instead, a *latching relay* can be used to keep the lights on after the control switch is released. Such an arrangement is shown in Fig. 9-12, where the relay has two sets of NO contacts. One set of contacts is used for the controlled circuit. The other set of contacts forms part of the control circuit, which con-

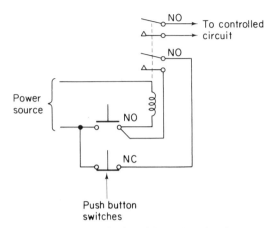

Figure 9-12 Typical latching relay circuit.

sists of the relay coil, power source, and two push-button switches, one NO and one NC. When the NO switch is pressed momentarily, current flows in the coil, and the controlled-circuit relay contacts make, applying power to the lights. The other set of relay contacts also make, completing the electrical circuit between the power source and coil, through the NC switch. The NO switch can then be released; the coil will remain energized, and the lights will remain on. When the lights are to be turned off, the NC push-button switch is pressed momentarily, power is remove to the relay coil, and both sets of the relay contacts break, removing power from the lights.

Time-delay relay. There are many control applications where it is desirable to have a relay become energized a short time after power is applied. An example of this is in a control system for an electrical generator. The generator is to be turned on immediately after the control switch is pressed, but the generator output is not to be connected to the load until the generator has had a chance to reach full operating speed. A conventional latching relay is used to control power to the generator, and a time-delay relay is used to control the output circuit of the generator.

There are a number of special-purpose time-delay relays used in control sytems. It is also possible to combine a conventional latching relay with a bimetallic temperature sensor (Sec. 6-2) to form a time-delay relay. Such an arrangement is shown in Fig. 9-13, where an electrical heater element is mounted near (or around) a bimetallic strip (which operates a set of NO contacts). When power is applied from the control circuit, current flows through the NC relay contacts to the heater. As the heater warms up, the bimetallic strip bends and eventually (after an adjustable time delay) closes the strip NO contacts, applying power to the relay coil. This closes the relay latching contacts (NO), and latches the relay so that the NO controlled cir-

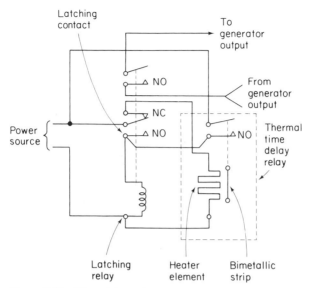

Figure 9-13 Time-delay relay combined with latching relay to form generator output control circuit.

cuit contacts make and remain closed. The relay NC contacts then open and remove power to the heater. The bimetallic strip returns to normal, and the strip contacts return to NO. The relay remains latched until power is removed, and the coil is deenergized.

Stepping relay. Figure 9-14 shows a *rotary stepping relay* used in many control systems. Each time the relay coil is energized, the armature actuates a ratchet wheel to advance one position. One or more cams may be attached to the ratchet, as shown. After a predetermined number of energizing pulses have been received by the relay coil, the cam rotates to a point where the spring leaf carrying the movable contact coincides with a notch in the cam. The end of the spring leaf drops into the notch, causing the contacts to make or break, thus closing or opening the controlled circuit. The next energizing pulse brings the leaf out of the notch and restores the contacts to their original position, where they remain until the next notch in the cam is reached. Each cam has its own spring leaf and contact set (and controlled circuit).

Meter relay. Figure 9-15 shows a *meter-actuated relay* (often called a *meter relay*). Such devices are used in some control applications where the control voltage or current is too small to energize the coil of a heavy-duty relay (that is required to handle a large current in the controlled circuit). As shown, there are three pointers on the meter. The center pointer is con-

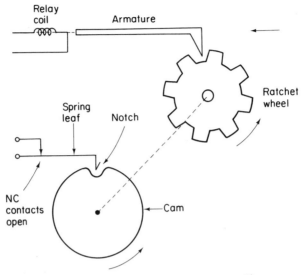

Figure 9-14 Rotary stepping relay operation.

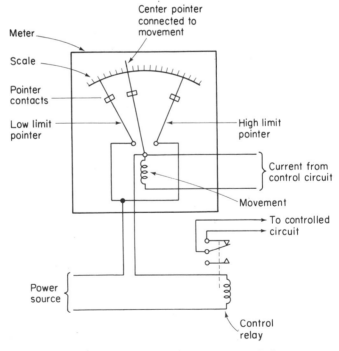

Figure 9-15 Meter-actuated relay (meter relay).

nected to the meter movement and moves over the scale in response to current from the control circuit. The other two pointers can be set to high and low limits of current. When the current reaches the point where the center pointer coincides with either of the two set pointers, the pointer contacts make, current is applied to the relay, and the controlled circuit is opened or closed.

A meter relay is sometimes used in control sytems where the controlled circuit operates valves or other actuators that control a process variable. The condition of the process variable is sensed by a transducer that produces a current output representing the condition, as discussed in Chapters 1 through 7. In this way, the current (or the process variable producing the current) may be kept within limits set by the meter pointers. For example, if the process variable increases, the transducer current increases, the meter movement pointer contacts the high limit pointer, the relay is energized, and the controlled circuit actuator is operated to decrease the process variable.

9-1.3 Electronic Relays

Electronic relays are used in control systems where the available control circuit current is not sufficient to operate a relay. A light-operated relay is a typical example. The output current of a photocell is typically a few microamperes, which is not sufficient to operate a heavy-duty relay. Instead, the current is amplified by using one or more transistors (Sec. 7-3.3). Such a combination of relay and amplifier is called an *electronic relay*. In addition to transistors, electronic relays use various other solid-state devices such as the SCR (silicon control rectifier), unijunction transistors, and light-actuated switches. Such devices are described in Sec. 9-4.

9-2 SOLENOIDS AND MOTORS

Solenoids and motors are two of the most common *actuators* used in control systems. As discussed in Chapter 1, an actuator is any device which converts a signal input to a mechanical motion. This is the reverse action of a transducer. The signal may be electrical, pneumatic, hydraulic, or mechanical. The motion, or output from the actuator, may be linear, rotary, reciprocating, etc. Gears and linkages may be used to change one type of motion to another. In general, solenoids are used to produce linear motion in response to an input signal from a controller, whereas motors are used for rotary motion. However, the reverse can be true in some cases.

9-2.1 Solenoid Actuators

Figure 9-16 shows the two basic types (pusher and puller) of solenoid actuators. Both the pusher and puller types consist essentially of a coil, a moving core, and a spring. When the coil is energized by a signal voltage from

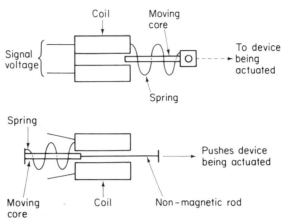

Figure 9-16 Basic pusher and puller solenoid actuators.

the controller or other source, the magnetic field around the coil pulls the core into the coil. When the coil is deenergized, the spring pulls the core out of the coil. The device to be actuated is attached to and moves with the core. Thus, the actuated device (such as a valve or cylinder) is pulled to one position as the coil is energized and restored to its original position when the coil is deenergized. For example, such a solenoid actuator can be used to open and close a valve. The spring which moves the core back to the deenergized position may be a part of the solenoid or part of the actuated device.

As shown in Fig. 9-16, a solenoid may also be used as a pusher. When the coil is energized, the core is moved in, and the nonmagnetic rod pushes the device to be actuated. When the coil is deenergized, the spring forces the core and the rod to their original positions. Such pushers frequently are used to remove objects from a moving conveyor belt upon a signal to the solenoid.

The solenoids described thus far are essentially ON-OFF (or two-position) linear actuators. These solenoids can be operated by either d-c or a-c power. When a-c power is used, the case and other metal parts around the coil are usually laminated to prevent solenoid "chatter" caused by the constantly changing magnetic fields. Linear solenoid actuators have many applications such as the operation of valves, brakes, gates, door openers, where mechanical force is required to push or pull an object in a straight line.

Figure 9-17 shows the operation of a rotary solenoid that produces rotary output strokes from about 5° to 90°. Rotary solenoids are often used to operate a ratchet connected to the lever arm of a rotary switch, where each output stroke of the solenoid advances the lever arm of the switch one position.

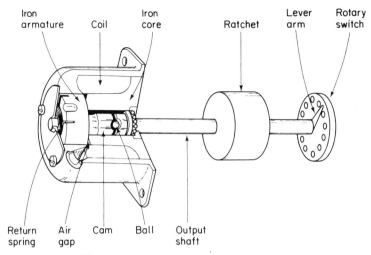

Iron armature | Coil | Iron core | Ratchet | Lever arm | Rotary switch

Return spring | Air gap | Cam | Ball | Output shaft

Figure 9-17 Rotary solenoid operation.

When current flows through the coil, a strong magnetic field is created between the iron armature and the iron core, which are separated by an air gap. The core is stationary. The armature is restrained from rotation and can move only in an axial direction. As a result of this pull, the armature is drawn across the air gap, carrying with it that half of the cylindrical cam to which it is attached. The other half of the cam is attached to the output shaft. The steel ball between the two halves of the cam acts like a cam follower. As the armature is drawn across the air gap, the armature half of the cam presses on the steel ball, forcing the steel ball against the output-shaft half of the cam. Since this shaft is restrained from axial motion by the bearing, the shaft is forced to rotate as the ball follows the contour of the cam. A return spring in the armature resets the armature to its original position when the coil is deenergized.

9-2.2 D-c Motor Actuators

The d-c motor is a common form of actuator. Figure 9-18 shows the basic elements found in a d-c motor, along with the symbols used for motors. Such a motor operates by the attraction between unlike magnetic poles and the repulsion between like poles. When current is passed through the *armature coil,* a magnetic field is set up around the armature. If current is passed through the armature coil in such a fashion that a north magnetic pole appears at the armature next to the north pole of the permanent magnet (known as the *field magnet*) the mutual repulsion between like poles causes the armature to rotate in a clockwise direction.

After a 180° rotation, the armature north pole is next to the field magnet

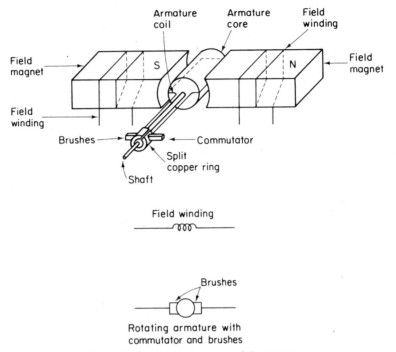

Field winding

⌐⌐⌐⌐⌐⌐⌐

Brushes

Rotating armature with
commutator and brushes

Figure 9-18   Basic elements of d-c motor.

south pole. At that point, the current flowing through the armature is
reversed by action of the *commutator*. In its simplest form shown in Fig.
9-18, a commutator consists of a split copper ring mounted on a rod. Each
half of the ring is insulated from the rod, and the halves are insulated from
each other. The commutator is mounted on the same shaft that is attached
to the armature and rotates with the armature and shaft. Each half of the
split commutator ring is connected to one end of the armature coil. Stationary *brushes* of copper or carbon make sliding contact with the commutator ring and apply current to the armature coil through the split ring.

Each time the armature (and the commutator ring which revolves with
the armature) makes a half-revolution, the direction of current through the
coil is reversed. Thus, the armature continues to rotate in one direction. By
connecting the device to be actuated to the motor shaft, the actuated device
is rotated in the same direction.

Practical d-c motors.   A practical motor is, of course, far more complex
than the simple device shown in Fig. 9-18. Although some small motors use
a permanent magnet as the field magnet, most larger motors use an electromagnet since a much stronger field can be obtained. The electromagnet is
shown by the *field winding* symbol in Fig. 9-18. Also, instead of using a
single loop of wire for the armature coil, many sets of loops are wound

around the armature core, each set connected to two opposite segments of the commutator ring. However, as shown in Fig. 9-18, the symbol shows only one set of brushes on the armature. Parallel sets of brushes are used for some heavy-duty motors.

Counter EMF.   As the armature winding cuts through a magnetic field set up by the field winding, a voltage is induced in the armature winding. (In effect, the motor becomes a generator of voltage. In fact, if an armature is rotated mechanically by an external force, the motor is a form of generator.) Since the voltage developed by the armature is opposed to the voltage in the power line supplying the motor, this voltage is called a *back* or *counter-electromotive force* (or CEMF).

When the motor is first started and there has not yet been time for the CEMF to build up, a relatively heavy current flows from the line. As the motor gathers speed and the CEMF builds up, the line current is reduced. Of course, the CEMF can never quite equal the line voltage, otherwise there will be no flow of line current and no rotation of the armature. The rotating effect, or *torque,* on the motor is proportional to the product of the armature current and the strength of the field. If the load on the motor is increased, the armature tends to slow down. As a result, the CEMF is reduced, which permits more line current to flow to the armature, and thus the torque is increased to meet the requirements of the increased load. If the load is decreased, the armature tends to speed up, the CEMF is increased, the current to the armature is decreased, and the torque decreases.

Thus, it is apparent that armature *speed* and CEMF are directly related. An increase in CEMF corresponds to an increase in speed and vice versa. Since the armature produces the CEMF, if the strength of the field is reduced, the motor will speed up to produce the necessary CEMF. When the field strength is increased, motor speed is reduced. Thus, motor speed is inversely proportional to field strength. Also, if line voltage is increased, the motor speeds up to develop a larger CEMF and balance the increased line voltage. The motor slows down when line voltage is decreased. Thus, motor speed (and torque) can be controlled by controlling either line voltage or field strength or both. Motor speed control is discussed further in Sec. 9-5.

Classes of d-c motors.   As shown in Fig. 9-19, there are three classes of d-c motors: *series, shunt,* and *compound.* The classification depends on the manner in which the field winding is connected to the armature.

In the series field motor, the field winding is connected in series with the armature winding and the power line. Since the two windings are in series, the same current flows through both windings, and the field winding has relatively few turns of wire (of the same size as the armature wire) compared to that of other motor classifications.

In the shunt field motor, the field winding is connected in *parallel,* or shunt, with the armature winding. The shunt field has fairly high resistance,

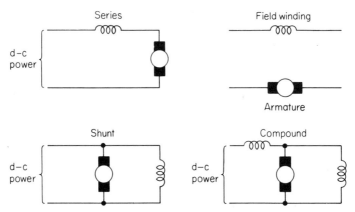

Figure 9-19 Classes of d-c motors (series, shunt, and compound).

so as to get maximum current flow through the armature. For this reason, the shunt field winding has many turns of relatively fine wire. Although little current flows through the field winding, proper field strength is maintained because of the many turns of wire.

In the compound field motor, two field windings are used, one series and one shunt. The series winding is wound with a few turns of heavy wire. The shunt winding is wound with many turns of fine wire.

Series motor characteristics.   Since the same current flows through the armature and field windings, a heavy current flows through both windings when a series motor first starts and there is little or no CEMF. This results in a *large starting torque.* As the motor reaches normal running speed, the CEMF buildup reduces the line current, and the torque drops to some normal value.

If the load is increased when a series motor is running, the motor tends to reduce speed. This produces a smaller CEMF, and the line current increases, resulting in a higher torque. The opposite occurs if the load is decreased when the series motor is running (speed and CEMF increase, line current and torque drop). If the load is completely removed from a series motor, speed increases sharply, producing a sharp increase in CEMF and a sharp reduction in line current. Field strength drops and speed increases to a point where the motor could fly apart or burn up if left running. For this reason, series motors are generally connected to the load by gears or other couplings that will present a load, even though the primary load may be removed.

Because of their high starting torque, series motors are often used in control applications where the inertia of a heavy load must be overcome. Typical series motor applications include electric locomotives and cranes.

Since the resistances of the armature and field windings of large series motors are quite small, there is a danger that the large starting current may be sufficient to burn out the windings. To overcome this problem, a resistor or other protective device is placed in series with the line to drop the starting voltage to some safe value. After the motor has reached its running speed and a sufficient CEMF is generated, the resistor is cut out. In very small series motors, the windings have fairly high resistances. Thus, a simple switch may be used to start such motors.

The simplest form of speed control for a series motor is a series rheostat, or variable resistance, as shown in Fig. 9-20. Speed is reduced when resistance is increased and vice versa. Once adjusted, the speed of a series motor remains fairly constant, provided the load is constant. However, an increasing load will reduce speed and vice versa. Series motor speed control is discussed further in Sec. 9-5.

Shunt motor characteristics. Because of a lack of CEMF, a heavy line current flows through the armature when a shunt motor is started (as in the series motor). However, since the shunt field winding has a high resistance, comparatively little current flows through the field. Thus, the starting torque of a shunt motor is less than that of a comparable series motor.

Since comparatively little of the line current passes through the shunt winding, the current flowing through the field varies only slightly with variations in line current. The strength of the field remains fairly constant under all conditions of load, and motor speed (which depends on field strength) also remains fairly constant. If the load is removed, the shunt motor tends to speed up but, since the field remains fairly constant, the faster the armature rotates, the greater the CEMF becomes. The shunt motor thus speeds up until the CEMF becomes equal to the line voltage. At that point, since the line voltage and CEMF cancel out, the armature current drops to nearly zero, and the motor cannot speed up further.

If the field winding of a shunt motor opens, due to burnout or other failure, the magnetic field and CEMF drop to near zero. As a result, armature current and speed rise sharply, possibly damaging the motor. As a precaution, devices such as overload circuit breakers or fuses usually are connected in the line so as to open when the current becomes excessive.

Shunt motors are generally used where constant speed is of greater importance than high starting torque. For that reason, shunt motors are used to operate machine tools and blowers. The simplest form of speed control for a shunt motor is a rheostat in series with the field winding or the armature as shown in Fig. 9-21. In the field winding rheostat system, an increase in resistance produces a weaker field and a higher speed. With the armature resistance system, increasing the resistance causes a reduction of armature voltage and a reduction of speed.

Compound motor characteristics. A compound motor has the characteristics of both the series and shunt motors. These include the ability to maintain a fairly constant speed with variations in load, and the ability to

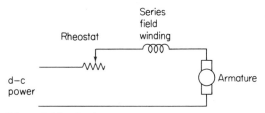

Figure 9-20 Basic speed control circuit for series motor.

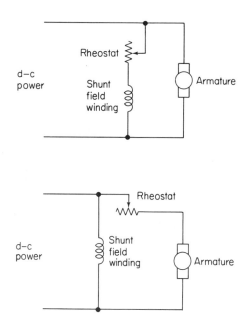

Figure 9-21 Basic speed control circuit for shunt motor.

produce a fairly large torque (both starting and running). Compound motor speed may be controlled by rheostats in the armature circuit, the field winding circuit, or both.

D-c motor reversal. A d-c motor rotates because of the repulsion between the magnetic poles of the field and similar poles of the armature. If the polarity of *both* field and armature windings are reversed, the motor will continue to rotate in the same direction. To reverse direction of rotation, the polarity of the field *or* the armature (but not both) must be reversed. Usually the armature polarity is reversed. Figure 9-22 shows a simple switch (DPDT) circuit for reversal of a d-c motor.

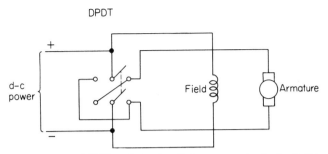

Figure 9-22 DPDT switch circuit for reversal of shunt d-c motor.

9-2.3 A-c Motor Actuators

A-c motors operate by the same basic principle as d-c motors (interaction between magnetic fields). However, the "field winding" of an a-c motor is called a *stator,* since it is stationary (generally in the form of a ring or cylinder). The "armature" of an a-c motor is called a *rotor* and is usually made up of metal segments, although some rotors have windings. There are three basic types of a-c motors: *induction, synchronous,* and *universal.* All three types operate by the principle of a *rotating magnetic field.* A-c motors can be operated with *single-phase* a-c power or *polyphase* (two-phase or three-phase) power.

Induction motor. Figure 9-23 shows the basic elements of an induction motor, which include a rotor and a stator. The metal rotor (consisting of copper segments or bars embedded in an insulated drum and connected at their ends to form a "winding") is driven by a revolving magnetic field produced in the stator windings when alternating currents are passed through the windings. As in the case of a d-c motor, unlike magnetic poles attract and like poles repel. If the two sets of stator poles shown in Fig. 9-23 are constantly changing (north to south and south to north) and the change *rotates around* the stator ring, the induced magnetic poles of the rotor are constantly repelled and attracted by the stator poles, so that the rotor spins in one direction.

Two (or more) electrical currents are required to produce the rotating magnetic fields. In a practical induction motor, the currents need not be from a separate source but can be two currents from the same source, one starting up a quarter-revolution (or 90°) behind the other. This is known as a *two-phase alternating current.* The graph of Fig. 9-23 illustrates what is meant by two-phase alternating current and how it relates to the two-phase induction motor.

At point 1 of the graph (90°) current *A* is positive maximum. The current flowing in stator winding *A* produces north (N) and south (S) poles as indicated. The heavy arrow shows the direction of the magnetic field. At point 1, current *B* is zero, and there is no magnetic field in stator winding *B*.

At point 2 of the graph (180°) current *A* is zero, and there is no magnetic field in stator winding *A*. However, current *B* is positive maximum, and the current in stator winding *B* produces north and south poles as indicated.

At point 3 of the graph (270°) current *B* is zero (no field in stator winding *B*), but current *A* is negative maximum, and the current in stator winding *A* produces north and south poles. However, because the current is now flowing in the opposite direction, the poles are in reverse position from those at position 1 (90°).

At point 4 of the graph (360°) current *A* is zero (no field in stator winding *A*), but current *B* is negative maximum, and the poles are in reverse position from those at position 2 (180°).

At each position of the graph, the north and south poles have, in effect, revolved in a clockwise direction. That is, the magnetic field (as indicated by the heavy arrows) has revolved as the currents proceed through their alternations. As the magnetic field of the stator revolves, so does the rotor. Note that in an induction motor, the rotor does not have a magnetic field of its own but depends on the magnetic field induced from the stator.

The speed of rotation (revolutions per minute) of an induction motor depends on the number of current reversals per minute divided by the number of *sets of poles*. With 60 Hz power there are 120 current reversals per second (7200 per minute). If there are two sets of poles as shown in Fig. 9-23, the field rotates at 3600 revolutions per minute (theoretically). In a practical induction motor with a load, the speed is about 3450 revolutions per minute (r/min).

The difference between synchronous speed and full load speeds can be explained as follows. If the rotor actually rotates at the same speed as the field, the rotor "winding" would never cut across the magnetic lines of force, and there is no torque. When a load is applied, the rotor tends to fall behind, or *slip* behind, the rotating field of the stator. The rotor then cuts lines of force, and torque is produced. The actual rotor speed slips behind the rotating field sufficiently so that it can induce enough current to produce the torque needed to satisfy the demands of the mechanical load. Slip can be measured in r/min, or as a percentage of the theoretical synchronous speed. When written as a decimal, the symbol for slip is *s*.

Torque is the "turning" or "twisting" force of a motor and is usually measured in pound-feet (or foot-pounds). Except when a motor is accelerating up to speed, the torque is related to the motor horsepower by the equation

$$\text{torque in pound-feet} = \frac{\text{hp} \times 5252}{\text{r/min}}$$

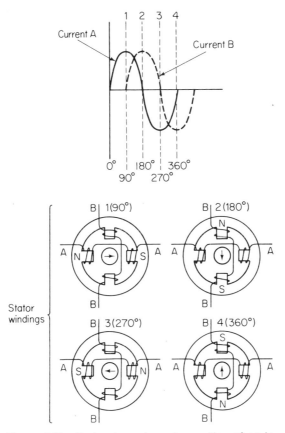

Figure 9-23 Basic elements and operation of an induction motor.

The torque of a 25 hp motor running at 1725 r/min can be computed as follows:

$$\text{torque} = \frac{25 \times 5252}{1725} = \text{approximately 76 lb-ft}$$

The most common, and simplest, rotor used in induction motors is the *squirrel-cage* type shown in Fig. 9-24. The rotor core consists of a laminated iron cylinder. Instead of wires, copper bars are inserted into slots in the surface of the core. The ends of the bars are joined together, thus forming a series of closed loops arranged in a sort of "squirrel cage" as shown. The magnetic field set up by the stator cuts across these closed loops, and large currents are induced in the loops. As a result of these induced currents the rotor becomes a magnet which is rotated by the revolving magnetic field of the stator.

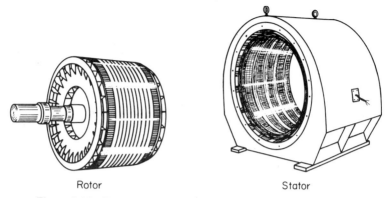

Rotor   Stator

Figure 9-24   Rotor and stator of squirrel-cage induction motor.

Operating induction motors from single-phase power.   When an induction motor must be operated from single-phase power, it is necessary to split the power into the equivalent of a two-phase current. There are several methods for doing this. The most common method is to use a *capacitor-start, split-phase induction motor* shown in Fig. 9-25. If a capacitor is placed in an electrical circuit, the phase of the voltage will be shifted from the phase of the voltage source. Thus, if two parallel circuits are connected across the line, one circuit containing a capacitor and the other having neither, a phase difference (of approximately 90°) will exist between the currents in both circuits. This is the equivalent of a two-phase current. The same results can be produced with an inductor instead of a capacitor; however, the capacitor is more popular.

In the circuit of Fig. 9-25, one stator winding (the *starting winding*) is wound on two opposite poles and has a capacitor in the circuit. The other stator winding (the *main* or *running winding*) is wound on the other two poles and has no capacitor in the circuit. Both circuits are connected in parallel across the single-phase power line. Because of the phase difference between the currents in both circuits, the motor starts as a two-phase induction motor. When the rotor reaches about 75% of full running speed, a centrifugal switch mounted on the rotor shaft opens the circuit of the starting winding. The motor then continues to run at normal speed, using the main winding.

Operating induction motors from three-phase power.   Induction motors are well suited to three-phase power. The effect on the motor is the same as if three generators are used, operating 120° apart electrically. The rotating field is provided by three stator windings wound on *three sets of poles,* 120° apart. These windings usually are placed in slots along the inner surface of the stator frame. Thus, the effect of "poles" is obtained without the use of protruding pole pieces. The result is a more compact motor with a reduced air gap between the stator and rotor.

236

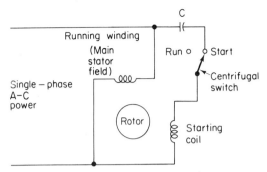

Figure 9-25 Capacitor-start, split-phase induction motor.

Three-phase induction motors may be large enough to deliver hundreds of horsepower. Such motors generally operate with line voltages of 220, 440, 550, or 2200 volts at 60 Hz. Three-phase motors may also be quite small. Such small motors are used in aircraft and missile control systems, where they operate from line voltages of about 200 volts at 400 Hz and produce outputs from a fraction to several horsepower.

Synchronous motor. The stator of a synchronous motor is essentially the same as that of an induction motor. However, the rotor of a synchronous motor does not depend on induced current from the stator for a magnetic field. In large synchronous motors, the rotor contains a winding (somewhat similar to that of the d-c motor) which is excited by a separate source of d-c current supplied by means of sliprings and brushes.

For starting purposes, copper bars are added around the rotor, forming a squirrel cage. A synchronous motor starts like an induction motor. When the rotor reaches near maximum speed, direct current is applied to the rotor winding. The magnetic poles created by the direct current lock in with the revolving field of the stator, and the rotor rotates with this field. Because the bars of the squirrel-cage portion of the rotor now rotate at the same speed as the revolving stator field, the bars do not cut any lines of force and have no induced current in them. In effect, the squirrel-cage portion of the rotor is removed from the operation of the motor.

Because the rotor has its own magnetic field, there is no need for the rotor to slip behind the rotating stator field. As a result, the rotor rotates in *exact step,* or *synchronization,* with the revolution of the field. In small synchronous motors, the rotor winding may be replaced by permanent magnets, eliminating the need for an external d-c source. Synchronous motor rotation speed is always synchronized (or *in sync*) with the rotating stator field which, in turn, is synchronized with the power line frequency. Thus, synchronous motor speed is constant, since power companies go to great pains to keep the frequency constant. For this reason, synchronous motors are used in such constant-speed applications as clocks and counters. Synchronous motors are controlled in the same way as induction motors.

Universal motor.   If alternating current is applied to a series d-c motor, the polarity of the magnetic fields around both the stator and rotor (field and armature) windings change in step with alternations of the current. Since like magnetic poles repel, regardless of whether they are two north poles or two south poles, a series d-c motor can be operated with a-c power (in addition to d-c power). Such an arrangement is known as a *universal motor.*

However, if a universal motor is to be operated with a-c power, certain precautions must be taken. The metal frame of the stator core and the metal rotor core must be laminated. Also, special care must be taken to reduce sparking at the brushes. Universal motors are frequently used for light-duty work such as in fans and vacuum cleaners.

Controlling a-c motors.   Even the simplest motors must have a means of starting and stopping, as well as overcurrent protection. The subject of motor control is quite complex, and is not described here due to space limitations. However, the following paragraphs cover the basics of a-c motor control. For a full discussion of a-c motor control, the reader's attention is invited to the author's *Handbook of Simplified Electrical Wiring* (1975, Prentice-Hall, Inc., Englewood Cliffs, N.J., 07632).

To reverse the direction of rotation of an induction motor, it is necessary to reverse the direction of the revolving magnetic field. In the two-phase motor, this is done by interchanging the connections to the terminals of either stator winding but not both windings. With three-phase power, the direction is reversed when all three windings are reversed. The term *plugging* is used when a motor running in one direction is momentarily reconnected to reverse its direction and is brought to rest very rapidly. *Jogging* (also known as *inching*) describes the repeated starting and stopping (but not reversing) of a motor at frequency intervals for short periods of time.

A-c motors, except for the universal type, are inherently constant-speed devices and do not lend themselves readily to speed control. Where such control is required, d-c motors operating from rectified a-c line current generally are used, as discussed in Sec. 9-5. Where the a-c motor must be used, speed control may be attained through the use of auxiliary mechanical devices such as gears and clutches. However, the speed of a-c motors can be controlled by solid-state devices as discussed in Sec. 9-5.

Several devices are used for electrical (nonsolid-state) control of a-c motors. These include *motor controllers, starters, switches, contactors,* and *relays.* These terms are often used interchangeably. Equally often, an incorrect term is used. For example, contactors and magnetic controllers are sometimes confused. For that reason, we summarize a-c motor control by providing some simple definitions of the devices used.

A *motor switch* provides only an ON-OFF function for a motor. However, some motor switches also provide for reversing direction of a motor.

A *motor controller* includes some or all of the following functions: starting, stopping, overload protection, overcurrent protection, reversing, changing speed, jogging, plugging, pilot-light indication, and possibly sequence control. Sequence control is particularly important in industrial control systems. Many industrial processes require a number of separate motors that must be started and stopped in a definite sequence such as a system of conveyors. When starting up, the delivery conveyor must start first with the other conveyors starting in sequence, to avoid pileup of material. When shutting down, the reverse sequence must be followed, with time delays between the shutdowns (except for emergency stops) so that no material is left on the conveyors. This is an example of simple sequence control (sometimes known as *interlocked control*). A motor controller can also provide control of auxiliary equipment such as brakes, clutches, solenoids, heaters, and signals. A motor controller may be used to control a single motor or a group of motors.

The terms *starter* and *controller* have virtually the same meaning. Strictly speaking, a starter is the simplest form of controller and is capable of starting and stopping a motor and providing it with overload protection. The starter shown in Fig. 9-26 can qualify as a controller, since it provides

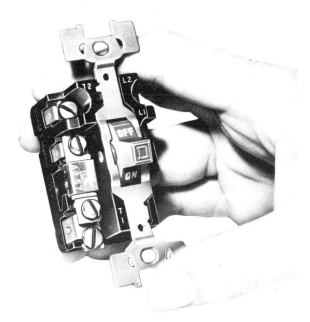

Figure 9-26 A-c motor starter with overload (overcurrent protection. (Courtesy of Square D Co.)

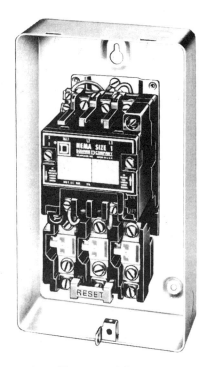

Figure 9-27   Controller for three-phase a-c motors. (Courtesy of Square D Co.)

an ON-OFF function (by means of a switch) and overload protection (by means of a bimetallic thermal overload element to the left of the switch). Such overload devices are essentially the same as the bimetallic thermal switches discussed in Sec. 6-2. The controller shown in Fig. 9-27 is typical of devices used to start and control three-phase a-c motors.

Motor controllers and starters can be either *manual* or *magnetic*. The term "manual" usually applies only to starters. However, both starters and controllers can be magnetic. The general classification *contactor* covers a type of magnetically operated device designed to handle relatively *high currents*. A conventional motor contactor is identical, in appearance, construction, and current-carrying ability, to a magnetic starter of equivalent size. The significant difference is that the contactor *does not provide overload protection*. A control *relay* is also a magnetically operated device, similar in operating characteristics to a contactor. However, a relay is used to switch low-current circuits.

9-2.4   Clutches and Brakes

Motors may be coupled to the device they actuate directly on the same shaft or through a train of gears. When gear trains are used, the actuated device may rotate faster or slower than the speed of the motor, depending on the *gear ratios* used. Also, the gear train can provide *torque amplification,* if required, since torque is increased in proportion to a decrease in speed.

Clutches.   In some cases, it is necessary to quickly engage and disengage a motor from the device being actuated, which is the function of a *clutch.* The most common type of motor clutch is the *friction disk clutch* shown in Fig. 9-28. The input shaft and clutch disk are rotated by the motor. Unless the clutch is actuated, the two disks are separated, and the output shaft (and the device being driven by the motor) does not rotate. When the clutch is actuated, the two disks are pressed together, and the output shaft (and the device being driven) is rotated because of the friction between the disk surfaces. When the disks are separated again (clutch not actuated), the output shaft and the device being driven are uncoupled from the input shaft (and motor), and they stop turning.

There are several methods for actuating clutch disks. In an automobile, the clutch is actuated by a mechanical linkage operated by a clutch pedal. In industrial control, the most common clutch actuator is an electromagnet or solenoid shown in Fig. 9-29. Current flowing through the solenoid coil winding produces a magnetic field that forces one of the clutch disks toward (or away from) the other clutch disk. In Fig. 9-29(a), the two disks are kept separate by the spring. Accordingly, the output shaft is uncoupled from the input shaft and does not rotate. Upon a signal in the form of an electric current sent to the solenoid, the two disks engage, and the output shaft rotates with the input shaft. When the signal is removed, the spring pulls the disks apart and the shafts are uncoupled. In Fig. 9-29(b), the disks are normally engaged, due to action of the spring. When a signal is sent to the solenoid, the disks are forced apart by the magnetic field, and the shafts are uncoupled. When the signal is removed, the spring forces the disks to engage once more, and the shafts are coupled.

Brakes.   The same principles used for clutches can also be used to brake a motor as shown in Fig. 9-30. Here, only one friction disk is mounted on the shaft between the motor and the actuated device. The other friction surface is supplied by the housing which encloses the brake. In Fig. 9-30(a), the

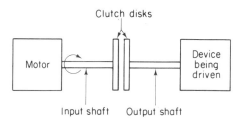

Figure 9-28   Basic friction disk clutch.

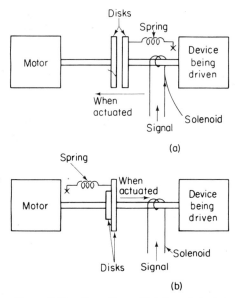

Figure 9-29 Operation of friction disk clutches actuated by an electromagnet or solenoid.

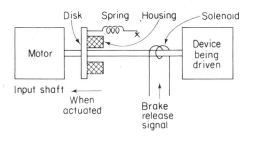

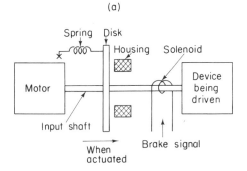

Figure 9-30 Operation of solenoid-actuated brakes.

spring holds the disk against the housing, thus preventing the shaft from rotating. When a *brake release* signal is applied to the coil, the disk is forced away from the housing, and the shaft is free to rotate. When the signal is removed, the spring pulls the disk back against the housing, and the rotation is stopped. In Fig. 9-30(b), the shaft normally is free to rotate, since the spring pulls the disk away from the housing. When a *brake* signal is applied, the disk is forced against the housing, thus stopping the rotation. When the signal is removed, the spring separates the disk from the housing, and the shaft is free to rotate again.

Combination clutch and brake.   The functions of clutch and brake can be incorporated in a single device. In the *brake-clutch* of Fig. 9-31(a), the output shaft normally is disengaged from the input shaft and is braked by action of the disk against the housing. On receiving a *start* signal, the brake is released, and the output shaft is coupled to the input shaft. In the *clutch-brake* of Fig. 9-31(b), the output shaft normally is coupled to the input shaft, and there is no braking. When a *stop* signal is applied, the disks are disengaged, and the output shaft is braked to a stop.

Variable-speed clutches.   The electromagnet clutches described thus far are ON-OFF devices. In many applications, it is desirable or necessary to vary the speed of the device being driven, while maintaining the motor at a

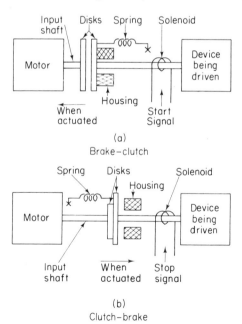

Figure 9-31 Operation of solenoid-actuated clutch-brakes.

constant speed. This is known as *proportional control* and is particularly effective with a-c motors, which are inherently constant-speed devices, where it is difficult to control speed. Proportional control can be obtained with variable-speed clutches.

The *eddy-current variable-speed clutch* shown in Fig. 9-32 uses a varying magnetic field to control speed. A metal drum is attached to the input shaft and revolves at the same speed as the motor. Inside the drum is a rotating pole assembly connected to the output shaft. There is no mechanical connection between the input and output. The stationary field coil mounted outside the drum is excited with direct current from a signal or control source. When the field coil is energized, a magnetic field is produced, cutting across the pole assembly and inducing a current in the assembly. This condition produces north and south poles, as shown.

When the drum rotates faster than the pole assembly, eddy currents are generated in the drum. These eddy currents establish their own magnetic fields and, by interaction with the magnetic fields of the field coil, produce a torque in the pole assembly (in the same direction as the rotation of the drum). (Note that eddy currents are the electrical currents produced in metal when the metal is cut by magnetic lines of force. In turn, eddy currents produce their own magnetic field.) The torque (which determines the speed of rotation for the pole assembly and the output shaft) is controlled by the excitation of the stationary field coil (the amount of current applied to the coil). By varying the amount of current to the field coil (say with a rheostat or potentiometer), the rotational speed of the device being driven can also be controlled, even though the motor speed remains constant.

The *magnetic-particle clutch,* shown in Fig. 9-33, is another device which can be used to control the output speed, depending on the magnitude of an external control signal (to provide proportional control). Here, a

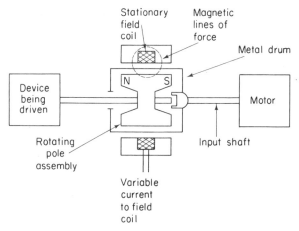

Figure 9-32 Eddy-current variable-speed clutch.

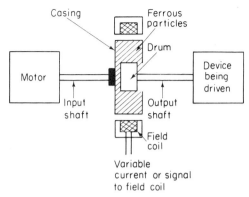

Figure 9-33 Magnetic-particle clutch.

drum attached to the output shaft is free to revolve within a casing. The casing is connected to the input shaft and revolves at the same speed as the motor. The space between the drum and casing is filled with ferrous particles, either dry or suspended in a fluid. The coupling between the drum and casing depends on the viscosity of the particles; the more viscous the mass of particles, the greater the coupling.

The viscosity of the particle mass is controlled by the magnetic field produced by an external coil. The stronger the field, the more the particles are magnetized and the more they cling together, thus making the mass more viscous. At maximum excitation of the coil, the particle mass acts as a solid, causing the drum and casing to revolve as one. At less than maximum excitation, the particles adhere to each other less firmly, and drum speed drops below that of the casing. Thus, by controlling excitation of the coil from the signal source, the degree of coupling between drum and case is varied. In turn, the amount of coupling determines the speed of the device being driven.

9-3 VALVES AND FLUID ACTUATORS

In this section we describe both valves and fluid actuators, since the two subjects are often interrelated.

9-3.1 Valves

A *valve* is a variable opening or orifice used to control the flow of a fluid or semifluid (such as powdered material). There are two basic types of valves: the *shut-off* type, where the opening is either completely open or completely closed, and the *throttling* type, where the opening may be adjusted to any size between the two extremes. Of course, a throttling valve does provide shut off when completely closed.

Valves can be manually operated (by handwheels, levers, etc.), operated by fluid actuators (as described in Sec. 9-3.2), or operated electrically by solenoids (Sec. 9-2.1). There is an infinite variety of valves used in modern control systems. Only the most commonly used are described here.

Plug-and-seat valves.   The *globe valve* shown in Fig. 9-34 is the most common type of plug-and-seat valve. Here, the flow of fluid through the valve is controlled by a movable *plug*. When the valve is open, fluid flows in from one port, through the valve, and out through the other port. As the plug is brought closer to the *seat* by action of the stem, the opening in the valve is reduced, as is flow through the valve. When the plug is completely seated, the opening is closed and the flow is shut off. Thus, the relative posi-

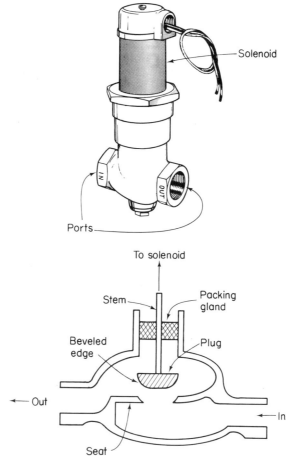

Figure 9-34 Solenoid-operated plug-and-seat (globe) valve.

tion between the plug and seat determines the effective opening (and flow) through the valve. The packing gland permits the stem to move up and down and yet forms a tight seal which prevents fluid from leaking past. The beveled edge of the plug forms a tight seal when the valve is closed. Frequently, hard rubber or fiber washers are attached to the bottom of the plug to improve the seal when the valve is closed.

The position of the plug is determined by the position of the stem. In turn, the stem position is determined by the actuator (hand, fluid, or electrical). Figure 9-34 shows the general physical appearance of a solenoid-operated plug-and-seat valve.

Slide valves.   The most basic type of slide valve is shown in Fig. 9-35. This valve is a *two-way,* or ON-OFF, type. When the actuator moves the valve rod or stem to the left, the valve is closed by the spool. Port A (intake port) and port B (exhaust port) are blocked off, and fluid flow through the valve is cut off. When the actuator moves the spool to the right, the fluid entering through port A flows around the narrow shaft of the spool and out through port B. In some valves, a spring (between the end of the spool and the left-hand end of the valve body) moves the spool to the right after the pressure exerted by the actuator is released.

Figure 9-36 shows a *three-way* slide valve used to operate a single-acting cylinder. When the spool is moved to the left, fluid under pressure enters through port A and exits through port C to the cylinder port. As a result of this pressure, the piston is forced to the left, compressing the spring and

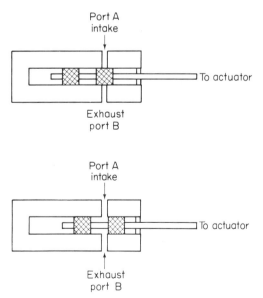

Figure 9-35   Two-way (ON-OFF) slide valve.

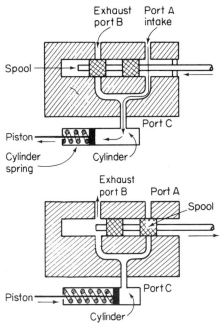

Figure 9-36 Three-way slide valve.

pushing the piston rod to the left. When the spool is moved to the right, port *A* is blocked, cutting off the fluid pressure. The cylinder spring forces the piston and rod to the right. The piston movement forces the fluid in the cylinder out through port *C*, through the valve, and out through port *B*.

Figure 9-37 shows a *four-way* slide valve used to operate a double-acting cylinder. When the spool is moved to the left, fluid under pressure applied through port *A* flows through the valve, exits at port *C*, and pushes the piston of the cylinder (and its piston rod) to the left. The movement of the piston forces any fluid remaining from the previous stroke out through port *D*, through the valve, and out through port *B*. When the spool is moved to the right, the fluid under pressure enters through port *A*, through the valve, and through port *D* to the cylinder. The piston and rod are pushed to the right, forcing the fluid remaining in the cylinder through port *C*, through the valve, and out through port *B*.

The advantage of using slide valves to manipulate the cylinders is that a relatively small force required to operate the valve can be used to control a much greater force exerted by the fluid on the cylinder.

Gate valve. The gate valve shown in Fig. 9-38 is used mostly for shutoff control. Here, a guided flat plate, or gate, operated by a linear actuator, controls the effective size of the passageway through the valve. If the flow is

248

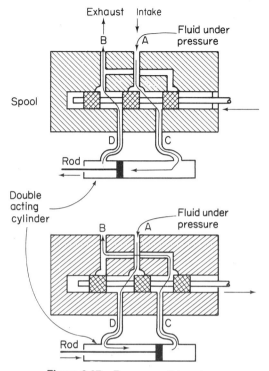

Figure 9-37 Four-way slide valve.

pulpy, such as wood pulp. the bottom of the gate may be sharpened to a knife edge. Thus, as the valve closes, suspended fibers in the fluid are sheared, ensuring a tight, clean closure without jamming.

Butterfly valve. The butterfly valve shown in Fig. 9-39 is useful if the line size is large and the line pressure low. The valve consists of a circular vane pivoted within the valve body. The position of the vane, and thus the effective size of the passageway through the valve, may be controlled by means of a rotary actuator or a linear actuator with a suitable mechanical linkage which converts the linear motion to rotary motion. A butterfly valve is especially suited for controlling the flow of pulpy or semi-solid materials which would foul plug- or slide-type valves. A butterfly valve is also used to control the flow of gas or air as, for example, at the air intake of an automobile carburetor.

Plug-cock (petcock) valve. The plug-cock valve shown in Fig. 9-40 consists of a tapered plug which fits snugly into a tapered hole in the valve body. When the valve is in the open position, a hole through the plug lines up with the passageway through the body, permitting fluid flow through the

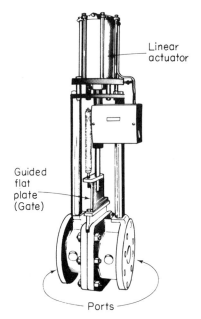

Figure 9-38 Gate valve.

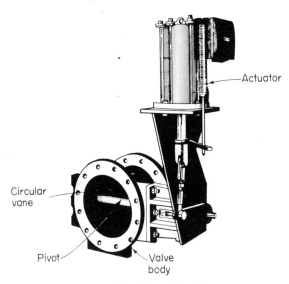

Figure 9-39 Butterfly valve.

valve. Rotating the stem 90° turns the plug so that its hole no longer lines up with the passageway through the body, thus cutting off flow. Plug-cock valves are used mostly for shutoff of flow. The stem may be operated by hand or by a rotary actuator.

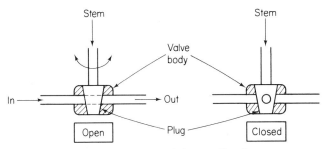

Figure 9-40 Plug-cock (petcock) valve.

9-4 ELECTRONIC CONTROL DEVICES

The use of electronic devices in control and instrumentation systems is constantly increasing. At one time, electron tube devices such as the diode, triode, thyratron, and ignatron were used extensively in control systems. Today, they have largely been replaced by solid-state electronic devices. For example, the electron tube diode and triode have been replaced by the solid-state diode and triode, discussed in Sec. 7-3.2 and 7-3.3, respectively. The thyratron and ignatron have been replaced by solid-state *controlled rectifiers* or *thyristors*. The subject of electronic control devices is quite extensive; it cannot be covered fully in one book, much less in this chapter. For that reason, we concentrate on the basics of a few selected solid-state control devices in this section. Some typical solid-state control device applications are discussed in Sec. 9-5.

9-4.1 Controlled Rectifiers and Thyristors

The controlled rectifier is similar to the basic diode (Sec. 7-3.2), with one specific exception. The controlled rectifier must be "triggered" or "turned on" by an external voltage source. The controlled rectifier has a high forward and reverse resistance (no current flow), without the trigger. When the trigger is applied, the forward resistance drops to zero (or very low), and a high forward current flows, as with the basic diode. The reverse current remains high, and no reverse current flows, so the controlled rectifier will rectify a-c power in the normal manner. As long as the forward voltage is applied, the forward current will continue to flow. The forward current will stop, and the controlled rectifier will "turn off," if the forward voltage is removed.

Of the numerous controlled rectifiers in use, many are actually the same type (or slightly modified versions) but manufactured under different trade names or designations. The term *thyristor* is applied to many controlled rectifiers. Technically, a thyristor is defined as any semiconductor switch whose bistable actions depend on PNPN regenerative feedback (Sec. 9-4.3). Thyristors can be two-, three-, or four-terminal devices, and both unidirectional and bidirectional devices are used in control systems.

9-4.2 Silicon or Semiconductor Controlled Rectifier (SCR)

With some manufacturers, the letters SCR refer to *semiconductor controlled rectifier* and can mean any type of solid-state controlled rectifier. However, SCR usually refers to *silicon controlled rectifiers*.

If four semiconductor materials, two P-type and two N-type, are arranged as shown in Fig. 9-41(a), the device can be considered as three diodes arranged alternately in series as shown in Fig. 9-41(b). Such a device acts as a conventional diode rectifier (Sec. 7-3.2) in the reverse direction, and as a combined electronic switch and series rectifier in the forward direction. Conduction in the forward direction can be controlled, or "gated," by operation of the switch.

Figure 9-42 shows the circuit symbol, block diagram, and basic physical construction of a typical SCR. Note that there are two basic arrangements for an SCR: one with the gate terminal connected to the cathode and one to the anode. The cathode gate is the most common arrangement.

SCR's are normally used to control alternating current but can be used to control direct current. Either a-c or d-c voltage can be used as the gate signal, provided that the gate voltage is large enough to trigger the SCR into the ON condition.

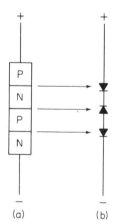

(a)   (b)

Figure 9-41 Semiconductor material arrangement and equivalent diode relationship of an SCR.

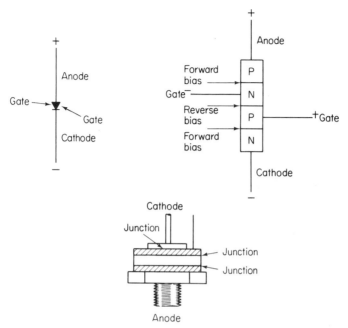

Figure 9-42 Symbol, block diagram, and basic construction of a typical SCR.

An SCR is used to best advantage when both the load and trigger are alternating current. With a-c power, control of the power applied to the load is determined by the *relative phase of the trigger signal versus the load voltage.* Because the trigger control is lost once the SCR is conducting, an a-c voltage at the load permits the trigger to regain control. Each alternation of alternating current through the load causes conduction to be interrupted (when the a-c voltage drops to zero between cycles), regardless of the polarity of the trigger signal.

Phase relationship between gate and load voltages. Figure 9-43 shows operation of an SCR with a-c voltages at the trigger circuit and load circuit. If the trigger voltage is in phase with the a-c power input signal as shown in Fig. 9-43(b), the SCR conducts for each successive positive alternation at the anode. When the trigger is positive-going at the same time as the load or anode voltage, load current starts to flow as soon as the load voltage reaches a value which will cause conduction. When the trigger voltage is negative-going, the load voltage is also negative-going, and conduction stops. The SCR acts as a half-wave rectifier in the normal manner.

If there is a 90° phase difference between the trigger voltage and load voltage (say load voltage lags trigger voltage by 90°) as shown in Fig. 9-43(c), the SCR does not start conducting until the trigger voltage swings

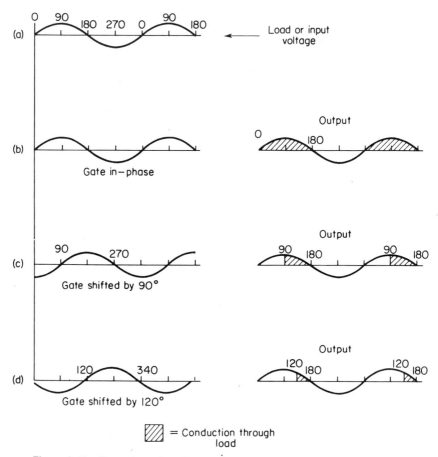

Figure 9-43 Operation of an SCR with a-c at both load and gate or trigger.

positive, even though the load voltage is initially positive. When the load voltage drops to zero, conduction stops, even though the trigger voltage is still positive.

If the phase shift is increased between trigger and load voltages as shown in Fig. 9-43(d), conduction time is even shorter, and less power is applied to the load circuit. By shifting the phase of the trigger voltage in relation to the load voltage, it is possible to vary the power output, even though the voltages are not changed in strength.

9-4.3 PNPN Controlled Switch or SCS

The PNPN controlled switch (often referred to as an *SCS* or *silicon controlled switch*) is similar in operation to an SCR. However, the SCS is a PNPN device with all four semiconductor regions made accessible by means of terminals. Figure 9-44 shows the circuit symbols, block diagram, and basic physical construction of a typical SCS. Note that two circuit symbols are used; both are in common use by manufacturers. Often, the SCS is used as an SCR with the extra gate terminal not connected.

For some control applications, the SCS can be considered as a transistor and diode in series. Figure 9-45 shows that arrangement. If a negative load voltage is applied to terminal 4, with a positive voltage at terminal 1, the SCS does not turn on, no matter what trigger signals are applied. However, with a positive voltage at terminal 4 and a negative voltage at terminal 1, the

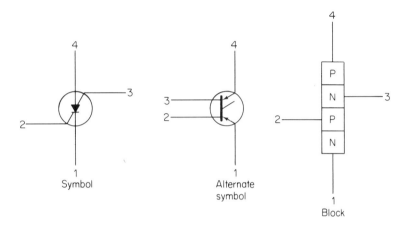

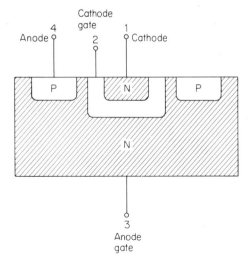

Figure 9-44 Circuit symbols, block diagram, and basic physical construction of typical SCS.

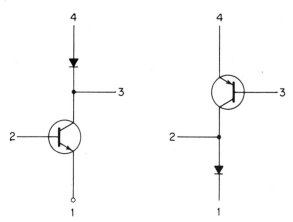

Figure 9-45 Equivalent diode-transistor representation of an SCS.

SCS is turned on by either a positive voltage at terminal 2 or a negative voltage at terminal 3.

The SCS also has the ability to turn off by means of a gate signal: that arrangement is shown in Fig. 9-46. Note that the gate turn-off method applies only when the conducting current is below a certain value. The SCS is often considered as two transistors (an NPN and a PNP) connected as shown in Fig. 9-47. Both transistors are connected so that the collector output of the NPN feeds into the base input of the PNP and vice versa. If a positive trigger voltage is applied to the NPN base, the NPN is turned on and some current flows in the NPN emitter-collector circuit. Since the NPN collector feeds the PNP base, the PNP is also turned on, and the PNP collector output feeds into the NPN base, adding to the trigger voltage.

Load current then flows through the two transistors from the NPN emitter (or rectifier cathode) to the PNP emitter (or rectifier anode). Load current continues to flow even though the trigger is removed, since the current flow also keeps both transistors turned on. The same condition can be produced by a negative trigger applied to the PNP base. This turns on the PNP which, in turn, turns on the NPN, until both transistors are fully conducting.

Normally, the currents do not stop until the load voltage is removed or, in the case of alternating current, the voltage drops to zero between cycles. However, if the load current is below a certain level (different for each type of SCS), the SCS can be turned off by a trigger voltage. For example, if a negative trigger voltage is applied to the NPN base, less current flows in the NPN emitter-collector circuit. As a result, the PNP is turned on less, and the PNP collector output drops to aid the turn-off trigger voltage at the NPN base. The feedback process continues until the load current is completely stopped.

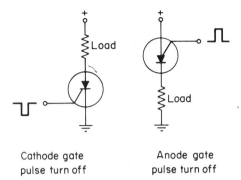

Cathode gate
pulse turn off

Anode gate
pulse turn off

Figure 9-46 Gate turn-off arrangement for SCS's.

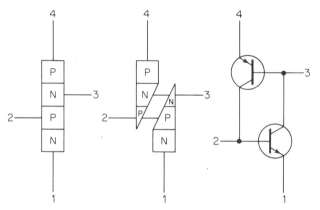

Figure 9-47 Equivalent two-transistor representation of an SCS.

9-4.4 Triacs and Diacs

The generally accepted meaning for the term *Triac* is *Bidirectional Triode Thyristor*. However, the term (as coined by General Electric) is used to identify a *tri*ode *a-c* semiconductor switch (or three-terminal switch). Like the SCR and SCS, the triac is triggered by a gate signal. Unlike either the SCR or SCS, the triac conducts in *both directions* and is, therefore, most useful for controlling devices operated by a-c power (such as a-c motors). Since an SCR or SCS is essentially a rectifier, two SCR's or SCS's must be connected back-to-back (in parallel or bridge) to control alternating current. (Otherwise, the alternating current will be rectified into direct current.) The use of two or more SCR's or SCS's requires elaborate control circuits (in many cases). The elaborate control circuits can be eliminated by a triac when a-c power is to be controlled.

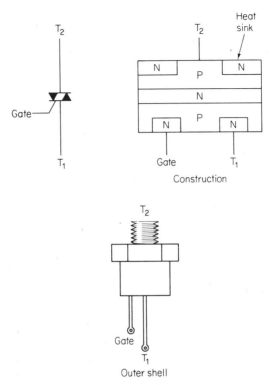

Figure 9-48 Symbol, outer shell, and construction of a typical Triac.

Figure 9-48 shows the circuit symbol and outward physical appearance of a Triac. Note that the symbol is essentially one SCR symbol combined with another complementary SCR symbol. Since the Triac is not a rectifier (when turned on and conducting, current flows in both directions), the terms "anode" and "cathode" do not apply. Instead, terminals are identified by number. Terminal $T1$ is the reference point for measurement of voltages and currents at the gate terminal and at terminal $T2$. The area between terminals $T1$ and $T2$ is essentially a PNPN switch is parallel with an NPNP switch.

Like the SCS and SCR, the Triac can be made to conduct when a "breakdown" or "breakover" voltage is applied across terminals $T1$ and $T2$, and when a trigger voltage is applied. Current continues to flow in one direction until that half-cycle of the a-c voltage (across $T1$ and $T2$) is complete. Current then flows in the opposite direction for the next half-cycle. The Triac does not conduct on either half-cycle unless a gate-trigger voltage is present during that half-cycle (unless the breakdown voltage is exceeded).

Triac trigger sources. Triacs can be triggered from many sources, as can SCR's and SCS's. One of the most common trigger sources is the *Diac* shown in Fig. 9-49. Note that there are two types of Diacs, diode and transistor. The term *Diac* (officially *Bidirectional Diode Thyristor*) was coined by General Electric to identify a *diode a-c* semiconductor device. The Diac is also known as a *bilateral trigger diode.* Whichever term is used, the Diac can be considered as a semiconductor device resembling a pair of diodes connected in complementary (parallel) form as shown in Fig. 9-49. The anode of the one diode is connected to the cathode of the other diode and vice versa.

Each diode passes current in one direction only, as in the case of a common diode (Sec. 7-3.2). However, the diodes in a Diac do not conduct in the forward direction until a certain "breakover" voltage is reached. For example, if a Diac is designed for a breakover voltage of 3 V, and the Diac is used in a circuit with less than 3 V, the diodes appear as a high resistance (no current flow). Both diodes conduct in their respective forward bias directions when the voltage is raised to any value over 3 V.

9-4.5 Unijunction Transistor and Zener Diode

The unijunction transistor (or UJT) is another commonly used trigger source for SCR's, SCS's, and triacs. Figure 9-50 shows a UJT connected in a basic trigger circuit. The UJT operates on an entirely different principle from that of the transistor (known as a bipolar or two-junction transistor) described in Sec. 7-3.3. The UJT is a *negative resistance* device. That is, under proper conditions, the input voltage or trigger signal can be decreased, yet the output or load current will increase. Once the UJT is "turned on," it will not "turn off" until the circuit is broken or the input voltage is removed. For this reason, the UJT makes an excellent trigger source for SCR's and other electronic control devices. When a small trigger voltage (either intermittant or constant) is applied, the UJT will "fire" and produce a large output voltage-pulse or signal that remains on until the circuit is broken (by switching off the base voltage).

As shown in Fig. 9-50, the UJT is a three-terminal device. The three terminals are designated *E* (emitter), *B*1 (base one), and *B*2 (base 2). The circuit of Fig. 9-50 is known as a *relaxation oscillator* and produces output

Diode type   Transistor type

Figure 9-49 Symbols for diode-type and transistor-type diacs.

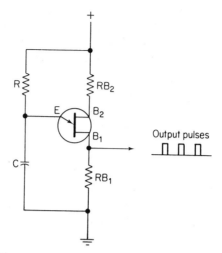

Figure 9-50 Basic UJT relaxation oscillator.

pulses (to be used as a trigger source for an SCR or other device) at a regular frequency or interval. The frequency or timing of this oscillator is controlled by the RC (resistor-capacitor) factor (or *time constant*) discussed in Sec. 8-4.3.

In the circuit of Fig. 9-50, capacitor C is charged through resistor R until the emitter voltage reaches a certain peak voltage, at which time the UJT turns on and discharges C through base resistor $RB1$. When the emitter voltage drops to a value of about 2 V, the emitter ceases to conduct, the UJT turns off, and the cycle is repeated. Thus, a series of trigger pulses appear at $B1$.

Zener diode. The zener diode, although not a trigger device, is often found in electronic control circuits such as the UJT trigger circuit of Fig. 9-51. Here, the zener diode is used to regulate the voltage applied to the UJT trigger circuit from the a-c power line. This action synchronizes the UJT with the power line. That is, if the a-c power line is the typical 60 Hz, the UJT will "fire" at 60 Hz and provide trigger pulses to the SCR at 60 Hz.

The zener diode functions primarily as a regulator. A zener is similar to the basic diode discussed in Sec. 7-3.2, except for an "avalanche" condition. Like a basic diode, the zener passes current in the forward condition and prevents current flow in the reverse condition. However, when sufficient reverse voltage is applied to a zener diode, the "avalanche" or "breakdown" condition occurs, and heavy reverse current flows. (Actually, any diode will operate similarly, but a zener is designed for this specific purpose.) With a zener, if the reverse voltage is increased above the breakdown point, additional reverse current flows, and the voltage is dropped through

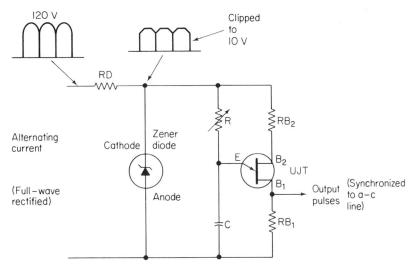

Figure 9-51 Zener diode used to regulate voltage applied to UJT trigger circuit (to synchronize UJT with power line).

a resistor (*RD* in Fig. 9-51) to a fixed level (known as the *zener voltage* or *zener level*).

In the circuit of Fig. 9-51, the 120 V line voltage is dropped to 10 V by the zener (110 V is dropped across resistor *RD*). In effect, the zener "clips" the peaks of the line voltage so that 10 V is applied to the UJT trigger circuit.

9-4.6 SUS and SBS

Figure 9-52 shows the symbols and equivalent circuits of the SUS and SBS. The SUS (silicon unilateral switch) is essentially a miniature SCR with an anode gate (instead of the usual cathode gate) and a built-in low-voltage avalanche (zener) diode between the gate and cathode. The SUS operates in a manner similar to the UJT. However, the SUS switches at a fixed voltage, determined by the internal zener diode, rather than by a fraction or percentage of the supply voltage. Also, the switching current of the SUS is generally higher than that of the UJT.

The SBS (silicon bilateral switch) is essentially two identical SUS structures arranged in an inverse-parallel circuit. Since the SBS operates as a switch with both polarities of applied voltage, the SBS is particularly useful for triggering Triacs with alternate positive and negative gate pulses.

9-4.7 Light Activated Control Devices

As discussed in Chapter 5, there are many types of devices available for converting light or radiant energy into electrical information. When this principle is used for electronic control devices, they are generally referred to

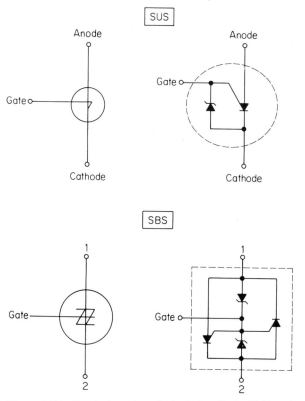

Figure 9-52 Symbols and equivalent circuits for SUS and SBS.

as *light activated semiconductors* (LAS). When light strikes silicon, there is an increase in current flow. In PNPN control devices, such as the SCR or SCS, this current increase acts as a trigger to turn on the device. The most common LAS devices are the LASCR (light activated SCR), LASCS (light activated SCS), and the opto-coupler (or light activated switch), all of which are discussed here.

LASCR and LASCS. Figure 9-53 shows the symbols and construction for the LASCR and LASCS. Note that the arrows indicate light or radiant energy striking the silicon chip. Both devices are similar in operation and characteristics to their corresponding SCR or SCS, except for the glass window on top of the can or enclosure. In a typical application the gate or trigger terminal is connected to a fixed voltage in the circuit. The LASCR or LASCS is then triggered by light passing through the window onto the silicon chip area.

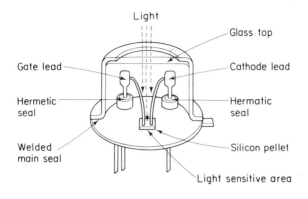

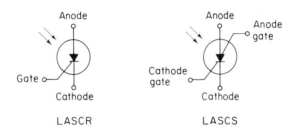

LASCR   LASCS

Figure 9-53 Symbols and construction for LASCR and LASCS.

Opto-coupler. Figure 9-54 shows a typical light activated switch or opto-coupler. As discussed in Chapter 5, when a photocell is exposed to very strong light, its internal resistance drops to zero (or near zero). This has the same effect as closing the contacts of a switch. When all light is removed from the photocell, its internal resistance increases to several million ohms. This has the same effect as opening the switch contacts.

In an opto-coupler, both the photocell and a light source are sealed in a light-proof enclosure. When the light source is off, the photocell resistance is very high, and the "switch" is "open". When the light source is on, the light strikes the photocell and causes its resistance to drop to zero. This "closes" the "switch". Such a switch is completely noise-free and spark-free, which are important factors for many control applications. The light source can be an incandescent lamp. However, most modern opto-couplers use a light-emitting diode (LED), similar to the LED's used for readouts, as discussed in Chapter 10.

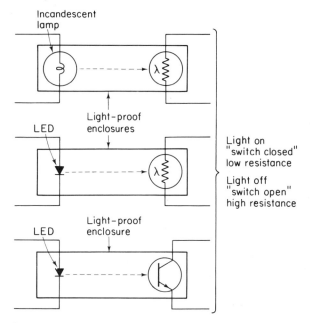

Figure 9-54 Typical opto-couplers (light activated switches, LAS).

9-5. TYPICAL ELECTRONIC CONTROL APPLICATIONS

We now describe some basic control applications for the electronic devices described in Sec. 9-4. Keep in mind that these represent only a very small part of the electronic control applications in present use.

9-5.1 Basic Phase Control

Phase control is the process of rapid ON-OFF switching that connects an a-c power supply to a load for a controlled fraction of each cycle—a highly efficient means of controlling average power to loads such as lamps, heaters, and motors. Control is accomplished by governing the *phase angle* of the a-c waveform at which the control device is triggered. The control device then conducts for the remainder of the half-cycle. Figure 9-55 shows the basic forms of a-c phase control.

The simplest form of phase control is the half-wave control shown in Fig. 9-55(a). This circuit uses one SCR for control of current flow in one direction only and is used for loads which require power control from zero to one-half of full-wave maximum. The circuit is also useful where direct current is required (or permitted) for the load.

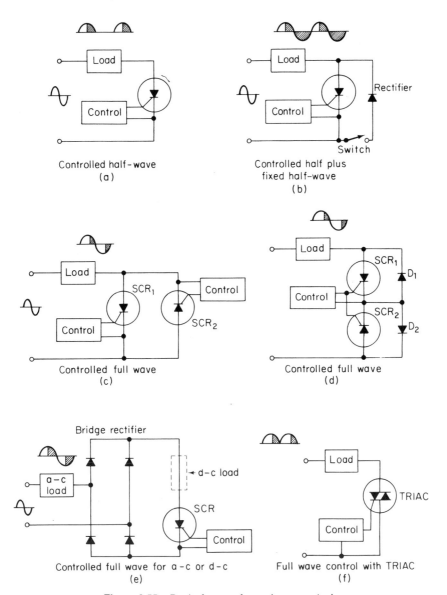

Figure 9-55 Basic forms of a-c phase control.

The addition of one rectifier, as shown in Fig. 9-55(b), provides a fixed half-cycle of power which shifts the power control range to half-power minimum and full-power maximum. The use of two SCR's, as shown in Fig. 9-55(c), controls from zero to full-power but requires two gate signals. A single gate or trigger signal is required for the zero to full-power control circuit of Fig. 9-55(d). The most flexible circuit, Fig. 9-55(e), uses one SCR inside a bridge rectifier and may be used for control of either a-c power or full-wave rectified d-c power.

Generally, the simplest, most efficient, and most reliable method for controlling a-c power is by the use of a Triac as shown in Fig. 9-55(f). This circuit provides full-wave control in both directions, using only one gate or trigger (often a Diac). For that reason, the Triac-Diac combination is, by far, the most popular a-c power control circuit.

Basic trigger circuits for phase control.   Most phase control trigger circuits use some form of relaxation oscillator such as the UJT oscillator discussed in Sec. 9-4.5. Since relaxation oscillators are simply *timing circuits,* they must be synchronized to an a-c power supply in order to provide proper phase control. That is, their timing cycles must start simultaneously with each alternation of a-c power. This synchronization is usually done by taking the oscillator input voltage from an a-c supply.

There are many ways of connecting the various versions of a basic oscillator circuit by using different triggering devices, control devices, supply, and load circuits. Figure 9-56 shows the basic half-wave and full-wave phase control circuits. With either circuit, the load, control device, and trigger device are all operated from the a-c power supply. Capacitor *C* is charged through adjustable resistor *R* on alternations of the a-c supply. In the half-wave circuit, *C* is charged once for each half-alternation or half-cycle.

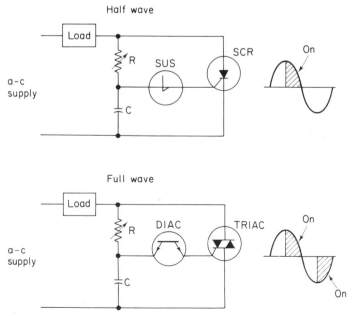

Figure 9-56   Basic half-wave and full-wave phase control circuits.

In the full-wave circuit, C is charged on both alternations of each cycle. When C is charged to a certain voltage, the trigger device turns on and provides a trigger signal to the control device. This switches the control device on for the remainder of the half-cycle, and power is applied to the load.

The time required to charge C to the trigger point is set by adjustment of R. In turn, the charge time determines the portion of the half-cycle in which the control device is on and the amount of power applied to the load. For example, assume that the value of R is adjusted to minimum resistance. This causes C to charge quickly and switch both the trigger and control devices on *early in the half-cycle.* Thus, maximum power is applied to the load. With R at maximum resistance, C charges slowly, and the devices switch on late in the half-cycle, providing minimum power to the load. Note that both trigger and control devices are switched off when the a-c power goes to zero between each half-cycle.

9-5.2 Static Switching Circuits

Electronic control devices are not limited to phase control. In fact, almost any static switching function performed by various mechanical and electromechanical switches can be accomplished with electronic controls, typified by the following two examples.

Electronic proximity switch. Figure 9-57 shows an electronic proximity switch. This circuit is ideally suited for use in elevators (both for door safety controls and floor selector buttons), supermarket door control, bank-safe monitors, flow switch actuators, and conveyor belt counting systems.

Capacitor $C1$, resistors $R1/R2$, and the sensor plate "capacitor" $C2$

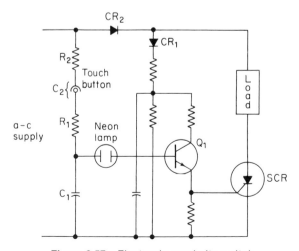

Figure 9-57 Electronic proximity switch.

form a voltage divider across the a-c supply. The voltage across $C1$ depends on the ratio of $C/C2$ and the supply voltage. The capacitance of $C2$, in turn, depends on the proximity to the sensor plate of any reasonably conductive and grounded object (metals, human body, etc.). As soon as the voltage across $C1$ exceeds the breakdown voltage of the neon lamp, capacitors $C1$ and $C2$ discharge through the base and emitter of transistor $Q1$. This discharge current is amplified by $Q1$ (as discussed in Sec. 7-3.3) and turns the SCR on to energize the load.

The load remains energized as long as the sensor plate or "button" is touched. When the button is no longer touched, the load is deenergized. A latching action can be provided by replacing $CR1$ with a short circuit (so that the SCR remains triggered on both half-cycles). The circuit must then be reset by an auxiliary contact in series with the SCR.

Electronic heater control. As discussed in Chapter 6, the mercury-in-glass technique can be used to sense and measure temperature and can be used as the basic element of a thermostat. One drawback of the mercury-in-glass system is its very low current handling capability, typically well below 1mA. This current is not sufficient to control a heater (to switch the heater on and off). The problem can be overcome by means of the circuit shown in Fig. 9-58. Here, an SCR is used to amplify the current of a mercury thermostat and to carry the heavy current required by the heater.

When the thermostat is open (low temperature), capacitor C is in the circuit and charges on each half-cycle of the a-c power. This triggers the SCR on each half-cycle and delivers power to the heater. When the thermostat is

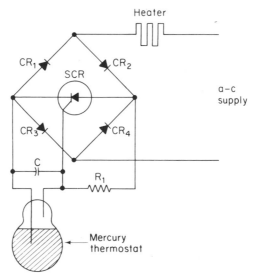

Figure 9-58 Electronic heater control.

closed (high temperature), capacitor C is shorted out of the circuit and cannot charge. The SCR can no longer trigger, and the heater shuts off.

9-5.3   Electronic Motor Control Circuits

One of the most common electronic control applications is *phase control of electric motor speed.* That is, the basic principles of phase control (Sec. 9-5.1) are used to control the speed of electric motors (Sec. 9-2.2). There are many motor control systems, so we discuss only the basic ones here. The main problem of motor speed control is *sensing motor speed* to provide *feedback* to offset changes in speed. If motor speed increases (due to a change in load or supply voltage) the feedback signal decreases motor speed and vice versa.

For d-c and universal motors, speed is usually sensed by monitoring the CEMF generated during the SCR on-time. For shunt and permanent-magnet motors, the CEMF is directly proportional to speed. If speed increases, the CEMF increases and vice versa. Thus, by comparing the CEMF against an adjustable voltage, and using that comparison to control SCR on-time, motor speed can be controlled within close tolerances. For series motors, the CEMF is almost proportional to speed, so the same basic principle can be used.

For a-c induction motors, speed is often sensed by means of a *tachometer generator* (Sec. 2-4). If speed increases, tachometer output voltage increases and vice versa. The tachometer feedback signal is used to control SCR on-time.

D-c and universal motor speed control.   Figure 9-59 shows the basic circuit for speed control of d-c and universal motors, using an SCR. This phase-control circuit operates by comparing the motor CEMF voltage against a reference voltage (which is rectified by $CR2$, and set by adjustment of $R2$).

If the CEMF is greater than the reference voltage (the motor is going faster than the selected speed), $CR1$ is reverse-biased (the cathode is more positive than the anode), the SCR is not triggered, and no power is applied to the motor. When the motor slows down, the CEMF voltage drops and is less than the reference voltage, $CR1$ is forward-biased (anode more positive), the SCR is triggered, and power is applied to the motor. The speed at which this occurs may be varied by changing the magnitude of the reference voltage (by adjustment of $R2$).

Induction motor control.   Figure 9-60 shows the basic circuit for speed control of a-c motors, using a tachometer generator geared to the motor. The output of the tachometer generator (or feedback signal) is rectified by diodes $D1$ through $D4$, and applied through resistor $R1$ to amplifier transistor $Q1$. This feedback signal is inverted by $Q1$ and is used to charge

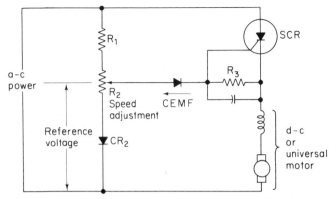

Figure 9-59 Basic d-c and universal motor speed control.

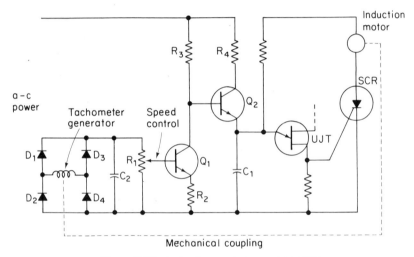

Figure 9-60 Induction motor speed control.

capacitor $C1$ through $Q2$. The charge on $C1$ determines the SCR on-time (as discussed in Sec. 9-5.1).

If the motor speed increases, the tachometer generator produces a greater output, which is rectified into a positive d-c voltage by the diodes. The positive voltage output is inverted by $Q1$ to a negative voltage which charges $C1$ to a lower value. The SCR then triggers at a later point, and the motor slows down. The charge on $C1$ (and thus the motor speed) can also be set by adjustment of $R1$.

9-5.4 Light-Actuated Control Applications

Light-presence detectors and light-absence detectors are typical examples of light-actuated control applications. Figure 9-61 shows the basic circuits for both applications.

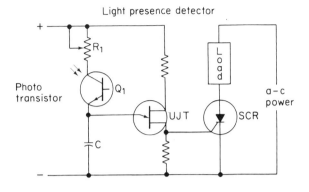

Light presence detector

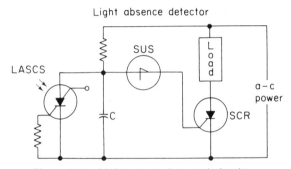

Light absence detector

Figure 9-61 Light-actuated control circuits.

For the *light-presence detector,* power to the load is switched on only when light is applied. Capacitor C starts to charge only when light strikes phototransistor $Q1$. The rate at which C charges is set by resistor $R1$. When C has charged sufficiently, the UJT fires and triggers the SCR, applying power to the load. When light is removed from $Q1$, capacitor C is disconnected from the power source and does not charge. Thus, the UJT and SCR do not trigger, and power is removed from the load. Typical applications of such a circuit are to activate warning or advertising signs at night by the headlights of approach vehicles.

For the *light-absence detector,* power to the load is switched on when light is interrupted. As long as light strikes the LASCS, capacitor C is shorted and does not charge. When light striking the LASCS is removed, capacitor C charges and the SUS anode goes positive on the next positive cycle of the power supply. This triggers the SUS and SCR, applying power to the load. When light is again applied to the LASCS, capacitor C is shorted and prevented from charging. Thus, the SUS and SCR do not trigger, and power is removed from the load. A typical application of such a circuit is to switch on house lights at dusk or whenever darkness occurs (such as on an overcast day).

9-6 MISCELLANEOUS CONTROL DEVICES

In addition to the electronic control devices described thus far, there are various mechanical and electromechanical controls in use. Many of these devices are used in existing control and instrumentation systems but will someday be replaced by more modern electronic controls. The four most common mechanical and electromechanical control devices include the motor-driven potentiometer, motor-driven variable transformer, saturable reactor or magnetic amplifier, and the servomechanism, all of which are described in the following sections.

9-6.1 Motor-driven Potentiometer

As discussed in Sec. 2-3, a potentiometer can be used as an angular-motion sensor. If a constant d-c voltage source is connected across the two outside terminals of a potentiometer, the voltage drop between the slider and one of the terminals depends on the degree of shaft rotation. Thus, a mechanical input signal to a potentiometer shaft produces a corresponding output signal voltage.

Where such a potentiometer is used for control purposes, an input signal may be applied manually. Where automatic control is required, the shaft may be rotated by some type of actuator in response to a signal applied to the actuator. For example, as shown in Fig. 9-62, a motor can be connected to the potentiometer shaft through gears (usually speed reducing gears). When a control signal is applied to the motor, the potentiometer shaft turns and produces an electrical output signal.

Motor-driven potentiometers are used where the output signal must be direct current. Such potentiometers are often used in *servo systems* as discussed in Chapter 10. The waveform of the output signal from the potentiometer may be altered to suit the needs of the control system as shown in Fig. 9-62. In some systems, several potentiometers are ganged together and operated by the same motor. Thus, a single signal to the motor produces several control signal outputs.

If a linear signal is desired, the resistance strip of the potentiometer is constructed so that the resistance varies uniformly over the entire length. Then, as shown in Fig. 9-62(a), the output voltage E varies linearly with shaft rotation. Note that in the circuit and graphs of Fig. 9-62, voltage output is indicated by E, and shaft angle by θ.

If the resistance strip of a nonlinear potentiometer is center-tapped and the circuit constructed as shown in Fig. 9-62(b), the output voltage will have a corresponding waveform. When the slider (and shaft) are at the zero angle position, the output is at maximum negative. As the slider is rotated in a clockwise direction, the output voltage becomes less negative, reaching zero when the slider is at the center-tap. As the slider is further rotated in a

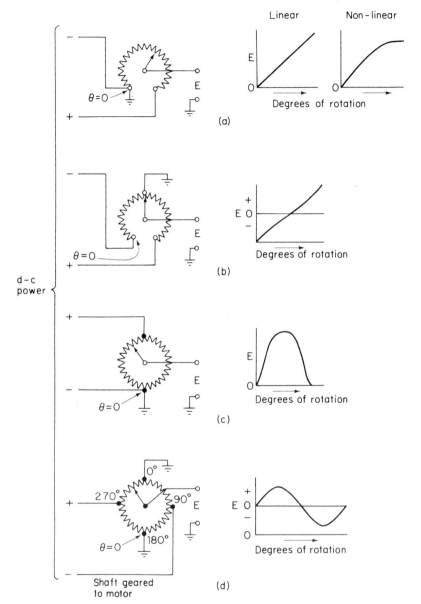

Figure 9-62 Output signals produced by various motor-driven potentiometer circuits.

clockwise direction, the output voltage becomes positive and increases as the rotation continues.

If the nonlinear resistance strip is shaped to form a continuous circle and connected as shown in Fig. 9-62(c), a corresponding waveform is produced.

273

Thus, the output voltage starts at zero, rises to a positive peak when the slider reaches the center-tap, and falls back to zero.

In Fig. 9-62(d), the nonlinear resistance strip is circular and tapped at four points (or quadrants), namely 0°, 90°, 180°, and 270°. The output voltage from such an arrangement is as shown in the corresponding waveform.

9-6.2 Motor-driven Variable Transformer

When a control system requires a variable a-c signal, a variable transformer may be used. Usually, an autotransformer with a single winding is used as shown in Fig. 9-63. The portion of the winding across the alternating source voltage E_{in} is the primary. The portion of the winding across the alternating output voltage E_{out} is the secondary. The ratio between the number of primary and secondary turns determines the output voltage. This ratio between the turns of the windings is determined by the position of the shaft-actuated slider. The shaft can be driven manually (with a knob), or can be motor-driven. As in the case of the motor-drive potentiometer, several variable transformers may be ganged together and operated by the same motor.

9-6.3 Saturable Reactor and Magnetic Amplifier

Saturable reactors and magnetic amplifiers operate by the principle of *variable reactance*. The basic circuit shown in Fig. 9-64 resembles an ordinary transformer with an input and an output winding. The *control winding* has many turns of relatively fine wire. The *output winding* has fewer turns of heavier wire. The output winding is connected in series with the load and a-c supply. The control winding has a variable resistance, switch, and a d-c source. Note that Fig. 9-64 also shows the standard electrical symbol for a saturable reactor.

When the switch is open, the control winding can be ignored. The device then acts as a simple iron-core reactor. Because the reactance is fairly high, a relatively small current will flow from the a-c source through the load.

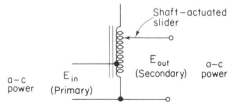

Figure 9-63 Motor-driven variable transformer.

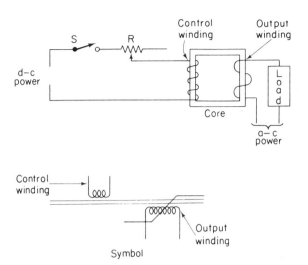

Figure 9-64 Saturable reactor and magnetic amplifier.

When the switch is closed, d-c power flowing through the control winding produces a steady magnetic field (or flux) which partially fills the core. Hence, less flux can be absorbed from the output winding before the core becomes saturated. As a result, the reactance of the output winding is reduced, and a larger current flows through the load.

The greater the amount of flux produced by the control winding, the less the amount of flux required from the output winding to drive the core to saturation and the lower the reactance of the output winding; thus, more current will flow through the load. Therefore, by varying the amount of flux produced by the control winding, the current flowing through the load and output winding are controlled. In turn, the amount of flux produced by the control winding depends, among other factors, on the amount of current flowing through the control winding from the d-c source. The current may be controlled by varying the resistance of R. Thus, the load current is set by adjustment of R. Where automatic control is required, R may be operated by an actuator (such as the motor-driven potentiometer discussed in Sec. 9-6.1).

Since the control winding contains many turns, a relatively small current flowing through the winding produces a flux sufficient to saturate the core. Thus, a small power input to the control winding will control a relatively large amount of power in the output winding. Since this is a form of amplification, the saturable reactor and the associated components are known as a *magnetic amplifier*.

As is characteristic of other amplifying devices, magnetic amplifiers are capable of controlling large amounts of power in response to small control signals. Thus, the very small outputs of photocells, thermocouples, and similar devices may be used to control thousands of watts of electrical power. A magnetic amplifier has greater resistance to mechanical shock, greater life expectancy, and is less vulnerable to temperature extremes than solid-state control devices. However, magnetic amplifiers have higher levels of distortion than solid-state controllers and are generally limited to low-frequency operation (generally at the power line frequency, 60 or possibly 400 Hz).

9-6.4 Servomechanisms

A servomechanism is a form of *continuous closed-loop control system* (Chapter 1). The basic elements of a servomechanism include a motor-actuator to operate some device and a control unit operated at some remote location. For example, assume that a large crane or arm mounted on a high tower must be operated from a control station on the ground. The arm must be turned and pointed in any given direction without the operator being able to see the arm. Figure 9-65 shows the most elementary form of ser-vomechanism used to accomplish this task. This device is known as a *null-balance servomechanism*.

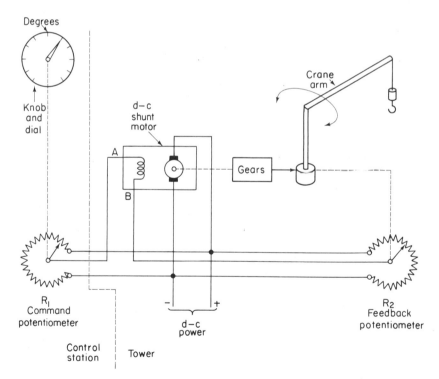

Figure 9-65   Basic null-balance servomechanism.

The arm is positioned by a shunt d-c motor (Sec. 9-2.2) through gears. A potentiometer $R2$ (known as the *feedback* potentiometer) is also geared to the arm and motor. Potentiometer $R2$, the arm, and the motor are all mounted on the tower and are connected to *command* potentiometer $R1$ at the control station through wires. Command potentiometer $R1$ is provided with a knob and dial marked off in degrees (or some other indication of arm position).

When the sliders of $R1$ and $R2$ are at equivalent positions, there is no voltage difference between them, and there will be no current flowing in the shunt field winding of the motor. When the slider of $R1$ is moved toward the positive side of the d-c supply (to turn the arm right) a voltage difference (known as the *error signal*) occurs between the two sliders, making point A of the field winding positive and point B negative. As a result, current flows in the field winding and the motor starts to rotate, moving the arm to the right. The farther the slider of $R1$ is moved from the original position (zero), the greater the error signal, and the faster the motor (and arm) rotates.

In addition to rotating the arm, the motor also moves the slider of $R2$ (through the gears). The direction of rotation of the motor is such that the slider of $R2$ is moved in the same direction as the slider of $R1$ was moved. When the slider of $R2$ reaches the equivalent of the new $R1$ slider position, the error signal is reduced to zero, current stops flowing in the motor field winding, and the motor, arm, and $R2$ slider stop.

When the slider of the $R1$ is moved toward the negative side of the d-c supply (to turn the arm left) an error signal is again created. However, the polarity of the error signal is now reversed, and current flows through the field winding in the opposite direction. This causes the motor to turn in the opposite direction, reversing the direction of rotation for the arm and direction of movement for the $R2$ slider. As before, the motor, arm, and $R2$ slider stop when the slider of $R2$ reaches a position equivalent to the new position of the $R1$ slider. Thus, the arm comes to rest at the position determined by the setting of the $R1$ slider (as indicated by the knob and calibrated dial).

Figure 9-66 shows a slightly more advanced form of null-balance servomechanism. Here, the error signal is an a-c voltage which is amplified by two transistors and applied to an a-c motor-actuator. Note that this a-c motor is called a *servomotor* on the illustration. Servomotors are similar to the two-phase induction motors described in Sec. 9-2.3 but with minor differences.

In a servomotor, the two windings (at right angles, or 90°) are called the *control winding* and the *reference winding*. The control winding is connected to the output of the *servo amplifier*. The reference winding is connected to the a-c line (generally operating at 60 or 400 Hz). If there is a

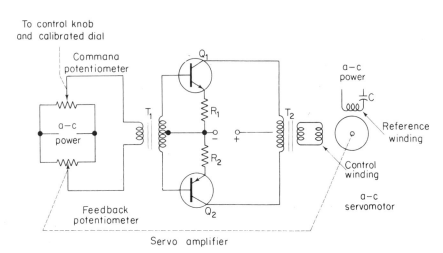

Figure 9-66 Simplified null-balance servomechanism with solid-state amplifier.

phase difference between the servo amplifier output (control winding) and the a-c line (reference winding), a revolving magnetic field is created, and the motor rotates. The direction of the magnetic field (and thus the direction of motor rotation) is determined by the *polarity of the phase difference* (whether the control voltage leads or lags the reference voltage). Since the reference voltage is fixed, the phase polarity is determined by the servo amplifier output in response to an error signal. If both voltages are in-phase or if the control voltage is completely removed from the winding, the motor stops.

An a-c error signal is generated when the slider of $R1$ is moved to a desired position (as indicated by the control knob and calibrated dial). This a-c error signal is applied to the amplifier through input transformer $T1$. The output of the amplifer (an amplified version of the error signal) is applied to the control winding of the servomotor through output transformer $T2$.

Note that the a-c line voltage is applied to the reference winding through capacitor C, which causes the voltage across the reference winding to lag behind the voltage of the a-c supply. When the voltage (from the servo amplifier output) applied to the control winding is in-phase with the supply voltage, the control voltage will lead the reference winding voltage, and the motor will rotate in one direction. When the control winding voltage is 180° out-of-phase with the supply voltage, the control voltage will lag behind the reference winding voltage, and the motor rotates in the opposite direction. When there is a zero error signal (the sliders of $R1$ and $R2$ are at the same position) there is no output to the control winding, and the motor remains stationary.

Like a d-c servomechanism, potentiometer $R2$ is geared to the servomotor (and whatever device is being driven). Thus, the motor, device being driven, and the $R2$ slider stop when the slider of $R2$ reaches a position equivalent to the selected position of the $R1$ slider.

10

BASIC INSTRUMENTATION DEVICES

As discussed in previous chapters, a primary sensing element senses the condition, state, or quantity of a process variable and produces a corresponding output signal. The output signal generally is one of three forms: a voltage signal (such as produced by a thermocouple), a resistance signal (such as produced by a strain gage), or a mechanical force signal. The force signal may be either of two types: a simple pressure signal (such as produced by a filled-bulb thermometer) or a high-pressure/low-pressure signal (such as produced by a venturi tube).

The signal is sent to a *controller,* a device which operates the final control element (such as a valve) in accordance with information contained in the signal. In simple systems, the controller may be either electrical or mechanical. In more complex, modern systems, the controller is often a computer or computer-like device. The signal may also be sent to an *indicator* such as a meter (if the signal is electrical) or to a pressure gage (if the signal indicates pressure). The indicator shows an instantaneous condition, state, or quantity of a process variable. The signal may also be sent to a *recorder,* which produces a continuous record of a process variable. In simple systems, the record is often printed on a chart. In computer-controlled systems, the record appears as a computer printout.

A primary sensor must, obviously, be located at the site of the process variable or condition being measured. It is frequently impractical, however, to place an indicator, recorder, or controller at a precise site. Instead, these devices are usually located at a central control station, where an operator may readily monitor a number of process variables simultaneously.

The output signal from a primary sensor is, therefore, sent to a control station. This is done by means of a *transmitter,* a device which converts a signal from a primary sensor to a form whereby the signal can be transmitted to a control station. In a sense, all the devices discussed thus far may be called *instrumentation.* However, in this book, we reserve that term for transmitters, indicators, recorders, and controllers, all of which are discussed in the following sections.

10-1 TRANSMITTERS

A transmitter consists of two sections. One section accepts an input signal from a process variable and measures the signal, generally by comparing the signal with an opposing reference voltage or force. The other section, as a result of that measurement, produces an output signal. Such output signals generally take one of two forms; an electrical signal, which may be transmitted over wires, or an air-pressure, or pneumatic, signal, which may be transmitted through metal tubing.

Transmitters may, therefore, be classified as producing either an electrical or pneumatic output signal. These two groups can be subdivided, depending on the nature of the input signal (electrical or mechanical force).

10-1.1 Electrical Transmitters

Electrical transmitters are generally classified as having a *voltage input, resistance input,* or *force input,* all of which are discussed here.

Electrical transmitter with voltage input. Figure 10-1 is a block diagram of a voltage input transmitter. The voltage input signal is measured by comparing it with a fixed reference voltage. Any difference voltage thus produced is fed to the control room through an amplifier. The signal is also fed back to the input circuit, where the signal acts to reduce the difference voltage to a very small value (or zero).

Figure 10-2 shows a typical voltage input circuit. Resistors $R1$ through $R5$ are arranged to form a bridge. Resistors $R1$ and $R2$ are of equal value and are much larger than the other resistors. Thus, the arms of the bridge form two paths of nearly equal resistance to the flow of current from the reference voltage source. The bridge is balanced by shorting the signal input terminals and adjusting $R5$ (the *zero adjust*) until the amplifier output is

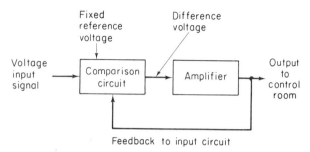

Figure 10-1 Block diagram of a voltage input transmitter.

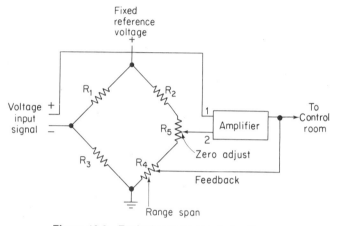

Figure 10-2 Typical voltage input circuit.

zero. In this condition, terminals 1 and 2 of the amplifier are at the same potential.

With the short removed and the input signal voltage applied, the bridge is unbalanced, and terminal 1 of the amplifier is made more positive by the additional voltage applied. As a result, the output current of the amplifier to the control room rises. An adjustable portion of the current, determined by the setting of *R*4 (the *range span*), is fed back to the bridge, rebalancing the bridge at the value of the signal input. Simultaneously, the voltage at terminal 2 is raised so that it approaches the voltage of terminal 1. (Note that the voltage of terminal 2 cannot rise unless there is a slight difference between the potentials at both terminals.) Thus, the output signal of the transmitter varies in direct proportion to the input signal.

Electrical transmitter with resistance input.   Figure 10-3 shows a modification of an input circuit to accommodate a resistance input such as produced by a wire strain gage. The circuit is essentially the same except that resistance *R*3 is a resistive sensor (e.g., a strain gage or resistance ther-

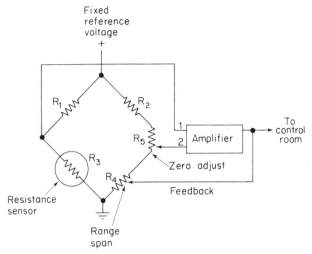

Figure 10-3 Electrical transmitter with resistance input.

mometer). The bridge is balanced by adjusting $R5$ until the amplifier output is zero (where the voltage at terminal 1 equals the terminal 2 voltage). When the resistance of the sensor $R3$ changes (as a result of a change in the process variable), the bridge is unbalanced and a difference in potential appears between terminals 1 and 2, thus relaying an output current from the transmitter to the control station. The current is also fed back to the bridge, tending to restore the balance.

Electrical transmitter with force input. Figure 10-4 shows an electrical transmitter designed to accept a force input. Here, the force is in the form of pressure applied to a Bourdon tube (Chapter 3). The motion of the tube resulting from the input pressure is applied as a force on a balance beam. The force is opposed by a force exerted by the input and zero springs. The difference between the two forces is force applied to the balance beam of a *force-balance system.*

As the input pressure increases, the upward motion of the Bourdon tube applies a force which moves up the end of the balance beam, thus widening the air gap between the ferrite pieces (iron strips) and a position detector (coil wound on an iron core). The resulting decrease in inductance causes an increase in the d-c output of the oscillator. When the input pressure decreases, the gap between the ferrite pieces and position detector also decreases, increasing the inductance and lowering the oscillator output signal. Thus, the oscillator acts as a sort of variable resistor, varying the output current in step with changes in input resistance.

The output current from the oscillator to the control station also flows through the coil of a magnet, which applies a balancing force on the other

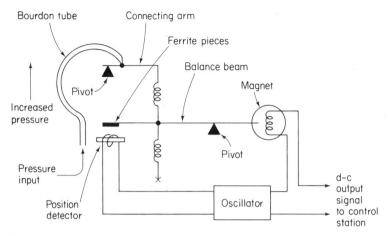

Figure 10-4 Electrical transmitter with force input (force-balance system).

end of the beam. The greater the pressure, the greater the output current and balancing force. The beam thus remains in constant balance.

10-1.2 Pneumatic Transmitters

The signals produced by electrical transmitters may be sent over a pair of wires. However, stray electrical and magnetic fields may induce currents in such wires, especially if they are long, which can interfere with the signals they carry. For this reason, many transmitters are of the pneumatic type, producing a signal consisting of air under varying pressure flowing through tubing. The variations in air pressure are proportional to the variations of the process variable.

Although there are many variations, the signal-producing portion of all pneumatic transmitters uses some type of *flapper-nozzle* arrangement. The position of the flapper (relative to the nozzle) is determined by the input to the transmitter. Accordingly, the pressure of the pneumatic output signal varies in proportion to the input from the primary sensor. By common agreement within the industry, the air-supply pressure is 20 psi (pounds per square inch). Also, the restriction and nozzle openings are designed so that, with the nozzle completely baffled by the flapper, the pressure of the pneumatic output signal is 15 psi. With the nozzle completely open, the pressure drops to 3 psi. Thus, if the transmitter is designed to send signals covering a temperature range from −100° to +800°, the pneumatic output signal at −100° has a pressure of 3 psi. At +800°, the pressure of the signal is 15 psi. Intermediate temperatures produce pneumatic signals at corresponding intermediate pressures.

Where the input to the transmitter is a voltage or resistance signal, the electrical transmitters discussed in Sec. 10-1.1 are used to convert the input signal to a d-c output signal. Then, that output is applied to an *electropneumatic transducer,* which converts the current to a proportional pneumatic output signal within the 3- to 15-psi range. Figure 10-5 shows the operation of a typical electropneumatic transducer. This device converts direct current to a force proportional to the current, thus operating a flapper by means of a structure similar to a dynamic loudspeaker. The coil is mounted over the pole piece of a permanent magnet. As current from the electrical transmitter section flows through the coil, a magnetic field (proportional to current) is formed around the coil. The field interacts with the field of the magnet and causes the coil to move in and out. A beam attached to the coil causes the baffle to move closer or farther from the nozzle.

The position of the baffle, relative to the nozzle, determines the pressure of the output signal. Thus, the signal pressure is proportional to the current flowing through the coil. In turn, the current is proportional to the signal received from the sensor.

10-2   RECORDERS

Modern industry requires a constant check on process variables to verify that they conform to predetermined values. In very simple systems, an operator can note and record readings at regular time intervals. But in a

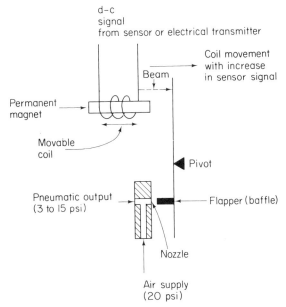

Figure 10-5 Electro pneumatic transducer using flapper nozzle to produce pneumatic output proportional to d-c input.

large plant where many process variables must be observed constantly, such as an oil refinery or nuclear generating plant, *automatic recorders* are used. These recorders monitor each variable and produce a continuous record of the values without the need for a human operator.

In some installations, a recorder is located near the site of a process variable. The input signal to the recorder may be the output from a primary sensor (usually a voltage, current, resistance, or pressure signal). If the recorder is located in a control room which is some distance from the process variable, a transmitter is used. The input signal to the recorder is then the output signal from the transmitter (usually a d-c signal or a pneumatic signal).

10-2.1 Recorder basics

Basically, a recorder consists of two sections. One is a measurement circuit which measures a signal received from a primary sensor or transmitter. The other section, acting upon information received from the measuring circuit, translates the information by actuating a pen, or some similar marking device, across a calibrated chart. The chart is moved by a timing device and thus provides a continuous record of the condition or value of a process variable.

The timing mechanism is a clockwork which moves a chart at a uniform rate. Generally, the clockwork is driven by a synchronous motor (Sec. 9-2.3). The marking mechanism is usually one or more *ink-fed pens,* although a heated stylus can also be used with specially treated paper. Most recorders also have an *indicating pointer* that move across a calibrated scale in step with a pen, thus showing the instantaneous value of a process variable.

10-2.2 Round-Chart and Strip-Chart Recorders

Figure 10-6 shows the two most common forms of recorders. The *round chart* consists of a circular paper disk with concentric circles ruled on the chart. The circles form the scale for the process variable to be recorded. In addition, equally spaced *time arcs* extending from the center to the rim are ruled on the chart. The time arcs have as their common center the point from which the pen arm is suspended. Thus, the pen moves along an arc when the chart is in motion. Since the chart is rotated at a uniform rate, the arcs become indicators of time. The pen also crosses the concentric circles when moving along the arc, thus recording the value of the process variable at the various time intervals. The pointer shows the instantaneous value of the process variable on the circular scale surrounding the chart.

The *strip chart* shown in Fig. 10-6 consists of a long roll of graph paper which is moved at a uniform rate beneath the pen. The measurement lines

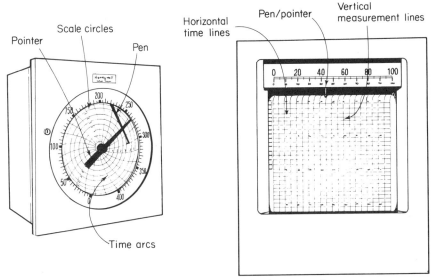

Figure 10-6 Round-chart and strip-chart recorders.

are straight parallel lines running lengthwise. The time lines are equally spaced horizontal lines. As the chart moves downward, the pen moves horizontally, thus constantly tracing a record of the process variable. The indicator pointer, which moves with the pen, indicates the instantaneous value of the process variable on the scale at the top. In some recorders, the chart moves horizontally and the pen/pointer move vertically.

For some recorders, the charts (and the indicator scales) are calibrated directly in units of the process variable (such as degrees of temperature, gallons per minute.) For other recorders, the charts and scales are calibrated in percentages of the range of values of the process variable. The recorder may be located near the process variable or in a control room at a remote location. Those recorders near the process variable may receive an input signal (voltage, current, etc.) directly from the primary sensor. Recorders in a control room receive signals sent by transmitters (electrical or pneumatic).

10-2.3 Typical Recorder Operation

There are many types of recorders, including *electrical, pneumatic,* and *those with alarms* (and *recorder-controllers* discussed in Sec. 10-4). Recorders can be further classified as to their method of operation, typically the *moving-coil* and *balance* type.

A *moving-coil* recorder is similar to the moving-coil meter described in Chapter 8. As current flows through the coil, it rotates, carrying with it a pen arm and pen. Springs oppose rotation of the coil. The greater the amplitude of the incoming signal, the more the coil rotates (and the more opposition applied by the springs). When these two opposing forces are equal, the pen comes to rest, having traced an indication of the process variable on the chart. Generally, the pen arm also has a pointer which indicates instantaneous values on a calibrated scale.

A *pneumatic* recorder is used where the input is a pressure signal. The measuring section of a pneumatic recorder is generally some form of pressure-to-motion transducer such as a Bourdon tube, bellows, or capsule (Chapter 2) to which a signal is applied. The resulting motion is linked to a mechanism which moves a pen across a chart or a pointer across a calibrated scale.

The recorder may be equipped with an *alarm mechanism* used to indicate when a predetermined high or low value of a process variable is reached. The indication may be a light or the sounding of a bell or buzzer. A typical alarm mechanism consists of one or more cams mounted on a pen-actuating device. The cam is so positioned that when a critical point is reached a cam follower falls into a notch on the cam, thus actuating a switch to which the cam follower is attached. An alarm sounds when the switch is closed.

10-2.4   Null-Balance Strip-Chart Recorder

Figure 10-7 shows the operation of a typical null-balance strip-chart recorder. This recorder consists essentially of a drive motor, chart paper, stylus or pen, servo motor, follow-up or feedback potentiometer, and amplifier. The drive motor pulls a roll of chart paper past the stylus at a uniform rate. Usually, the chart paper is marked in units of time so that the recorder provides a permanent time-related record (or log) of the signals being measured. The stylus is moved across the chart paper (at right angles to the chart paper lines of motion) by the servo, which, in turn, receives an input voltage from the amplifier.

The direction of servo motor (and stylus) motion is determined by the input signal polarity, whereas the amplitude of motion is determined by signal amplitude. As the stylus moves, the follow-up potentiometer produces an error signal or voltage proportional to the stylus position. The follow-up potentiometer voltage is of opposite polarity to that of the input signal. Therefore, when the stylus has moved to a position corresponding to the input signal, the follow-up potentiometer output cancels the input signal, stopping the stylus. Thus, the stylus maintains a position corresponding to the input signal and traces out a line on the chart paper that corresponds to the input.

Such strip-chart recorders are particularly useful for recording values over a long period of time, such as monitoring industrial processes. For example, assume that a pressure must be monitored over a 24-hour period.

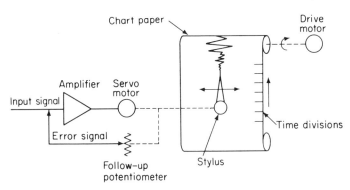

Figure 10-7 Operation of typical null-balance strip-chart recorder.

The pressure is converted to a voltage by a transducer (Chapter 3). The voltage is fed to a recorder amplifier input, and a stylus moves to a position corresponding to the voltage (wich represents the pressure). A drive motor is started, and the time (and/or date) is recorded on the paper or is noted. The recorder maintains a constant record of pressure. That record can be retained for future use.

10-3 SYNCHROS, SELSYNS AND AUTOSYNS

The *synchro* (also known as a *selsyn* or *autosyn*) is a generic name for a group of rotating devices which resemble motors in that they contain a rotating armature (rotor) within a fixed set of field windings (stator) as shown in Fig. 10-8. Synchros are generally used in pairs for transmitter systems which form a link between control and instrumentation functions. Typically, when the rotor of one synchro (called the *master unit, synchro generator,* or *transmitter*) is rotated a certain number of degrees, an electrical signal is generated. The signal is relayed by wires to another synchro (called the *slave unit, synchro motor,* or *receiver*), and causes the rotor of the receiver synchro to rotate the same number of degrees as the transmitter. In a typical system, the transmitter synchro shaft is geared to a control function, with the receiver shaft attached to an indicator (pointer, dial, etc.). The indicator then follows the control function.

The output torque of a synchro is relatively small, and a receiver synchro is limited to light loads such as moving a pointer across a scale to indicate the angular displacement of some device operating a transmitter synchro. Synchros are sometimes used with an amplifier to actuate heavier loads. Because the two synchros are connected by electrical wires, the actuating (transmitter) and actuated (receiver) may be a considerable distance apart.

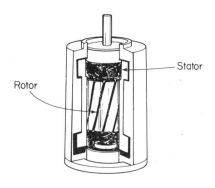

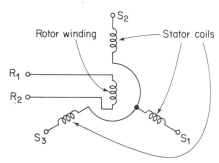

Figure 10-8 Construction and diagram of typical synchro.

As shown in Fig. 10-8, the stator winding consists of three coils 120° apart, whereas the rotor is a single winding. By convention, the rotor terminals are labeled $R1$ and $R2$; the stator terminals are labeled $S1$, $S2$, and $S3$.

A basic form of synchro system uses two synchros, one transmitter and one receiver. There are other combinations, such as using two transmitters and a differential receiver, or using a differential transmitter. We now describe a typical one-transmitter, one-receiver system.

10-3.1 Typical Synchro System

For purposes of illustration, a synchro transmitter is a form of *variable transformer,* where the coupling between primary and secondary may be varied by rotation of a rotor. Figure 10-9 shows the relationship between coupling and induced voltage in a transformer. When two coils are magnetically coupled, the magnetic field produced by the current induced in one coil opposes (in polarity) the magnetic field of the other coil. The degree

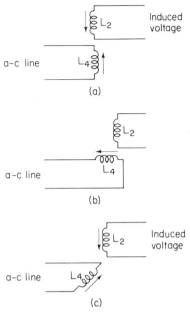

Figure 10-9 Relationship between coupling of rotor and stator synchro coils and induced output voltage.

of magnetic coupling, and thus the magnitude of the induced current depends on the *relative positions* of the coupled windings (assuming that there is a closed circuit through which current may flow).

For example, as shown in Fig. 10-9(a) the plane of $L4$ is in line with the plane of $L2$, maximum current flows through $L2$. (The arrows show the polarities of their respective magnetic fields. The arrowhead indicates a north pole.) If the plane of $L4$ is at right angles to the plane of $L2$ as shown in Fig. 10-9(b), minimum current flows in $L2$. (The absence of the arrow indicates that there is no magnetic field around $L2$.) If the plane of $L4$ is at some intermediate position, as shown in Fig. 10-9(c), the induced current in $L2$ is also at some corresponding intermediate value.

Figure 10-10 shows the relationship of induced current and coupling in a synchro transmitter. As shown in Fig. 10-10(a), the rotor is lined up in the same plane as the $L2$ coil. (In that position, a synchro transmitter or motor is said to be at a *zero degree position*). Maximum voltage is induced in the $L2$ coil, and maximum induced current flows through $L2$ (assuming that the coil is part of a closed circuit). The angular position and polarity of the magnetic field are indicated by the arrows at the side of the winding.

Intermediate values of induced current flow through coils $L1$ and $L3$. Because both coils occupy the same angular relationship to the position of

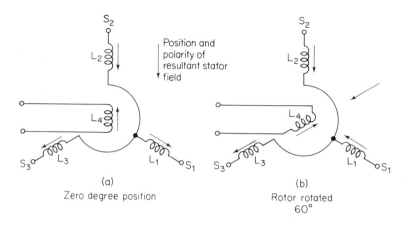

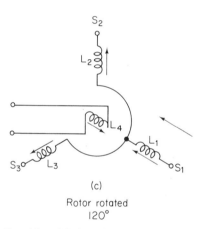

Figure 10-10 Relationship of induced current and coupling in synchro system.

the rotor, the voltage induced in $L1$ is equal to that induced in $L3$. The positions and polarities of the respective magnetic fields are indicated by the arrows. The resultant magnetic field of all three stator coils is indicated by the heavy arrow. Note that the resultant field assumes the same angular position as that of the rotor's magnetic field but opposed in polarity.

In Fig. 10-10(b), the rotor has been rotated 60° in a clockwise direction. The stator's resultant field is indicated by the heavy arrow. The resultant field again opposes the polarity of the rotor's magnetic field, although the field has rotated 60° in a clockwise direction (in step with the rotor).

In Fig. 10-10(c), the rotor has been rotated 120° in a clockwise direction from the zero degree position. Again, the resultant magnetic field of the stator rotates with the rotor, although still opposed in polarity. Thus, the magnetic field of the stator always assumes the same angular position of the rotor's magnetic field, though always opposed in polarity.

Figure 10-11 shows a synchro transmitter connected to a synchro receiver. Both rotor windings are connected in parallel and to the a-c power line. Hence, the polarities of the rotor magnetic fields are similar. The external lead of the transmitter $L1$ connects to the corresponding receiver $L1$ lead, as do $L2$ and $L3$. Because they are in series, the same current flows through the transmitter and receiver $L2$ windings. Similarly, the same current flows through the corresponding $L1$ and $L3$ windings of the transmitter and receiver.

Because the windings are identical, each receiver field coil has the same magnetic field as the corresponding transmitter field coil. However, since the direction of current flow through the receiver coils is opposite to that of the transmitter coils, the magnetic fields are opposed. Thus, although the receiver's resultant stator field assumes the same angular position as the transmitter's stator field, the two fields are always opposed in polarity.

Summary of synchro system operation. The following summarizes the relationship between transmitter and receiver synchros.

1. The magnetic field around the transmitter rotor has the same polarity as the field around the receiver rotor.

2. The resultant stator field of the transmitter is always opposed in polarity to the rotor field and rotates in exact step with rotation of the rotor.

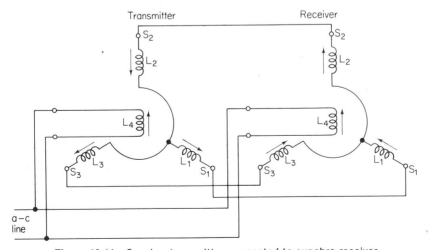

Figure 10-11   Synchro transmitter connected to synchro receiver.

3. The resultant stator field of the receiver is opposite in polarity to the resultant stator field of the transmitter and rotates in step with the transmitter's stator field.

4. The magnetic fields of both the rotor and stator of the receiver are always in the same plane and polarity. Hence, they line up and the receiver's rotor revolves in step with rotation of the stator field.

5. Thus, as the transmitter's rotor is turned, the receiver's rotor is revolved to the same degree and in the same direction (as is any device attached to the receiver's shaft).

10-4 CONTROLLERS

A controller is the brain of an automatic control system. As discussed in previous chapters, the primary sensing element produces a signal which corresponds to the *instantaneous* value of the process variable. The signal is accepted by the controller, which measures and compares the signal with a *set-point* signal (representing the desired process variable value). If there is any difference (or *deviation*) between the two signals, the controller sends an output signal (electrical, pneumatic, or possibly mechanical) to the final control element (valve, etc.). The control element then acts to bring the process variable value to the set-point value, thus eliminating any deviation.

When a controller acts to maintain the condition of a process variable within certain limits above and below a set-point value, it is known as an ON-OFF, or *two-position controller.* There are also *proportional controllers,* which provide graduated control of a process variable. For example, assume that a controller is used to operate a valve that sets the temperature of a liquid. When the valve is open, the temperature rises and

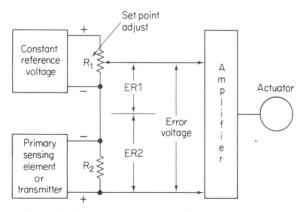

Figure 10-12 Basic principle of electrical controller.

vice versa. Instead of manipulating the valve so that it is either completely open or completely closed, the proportional controller produces a throttling action, where the valve opening is reduced as temperature rises and is opened wider as temperature is reduced. With the proportional controller, the degree of valve opening is dependent on the amount of deviation between the process variable value and the set-point value.

All controllers consist of two main sections. One is a *measurement section,* which measures the value of an incoming signal received from a primary sensor or transmitter and compares the signal with a set-point signal value. The other is a *control section* which, acting upon deviation between the two signals, sends an output signal to a final control element (to reduce and eliminate the deviation). Note that the two sections are not always clearly defined in all controllers.

There are three basic types of controllers: the *nonindicating* or *blind controller,* the *indicating controller,* and the *recorder-controller.* A blind controller generally consists only of the measurement and control sections and is usually an ON-OFF controller. An indicating controller contains, in addition to the two basic sections, an indicating device which shows the instantaneous value of a process variable. A recorder-controller consists of a recorder which, as discussed in Sec. 10-2, contains a measurement section and a control section for operation on information received from the measurement section. The controller may be located near the process variable or in a control room at a remote location. A nonindicating controller is generally located near the process variable. A recorder-controller is usually located at the control station. An indicating controller may be installed at either site.

10-4.1 Electrical Controllers

The basic principle of an electrical controller is shown in Fig. 10-12. The electrical input signal from the primary sensing element or a transmitter flows through $R2$, producing a voltage drop $ER2$ which corresponds to the value of the process variable. A constant reference voltage $ER1$ opposes this voltage. The *error voltage* is the difference between $ER1$ and $ER2$.

The *set point* is determined by adjustment of $R1$ to fix the value of the reference voltage $ER1$. Since $ER1$ is constant (once adjusted to the set point) and $ER2$ corresponds to the process variable, the error voltage corresponds to the variable. As shown, the error voltage is usually amplified before it is applied to the final control element (valve, actuator, etc.).

There is an infinite variety of electrical controllers, many of which are being supplanted by computers or computer-like devices such as microprocessors. Thus, we shall not attempt to describe all of them here. Instead, we concentrate on a simple system which illustrates the principle of *floating control.*

Floating control system. Figure 10-13 shows the elements of a basic floating control system used to control the temperature of a liquid in a tank. The liquid is heated by steam flowing through a jacket surrounding the tank. The steam is controlled by a valve which, in turn, is opened and closed by a motor actuator. The controller receives an input signal from a thermocouple mounted inside the tank and produces a corresponding output signal to the motor actuator. The control element is an indicating meter, the moving coil of which is operated by signal voltages from the thermocouple. The meter indicator pointer makes contact with two sets of contacts. One set of contacts is at the low limit of the process variable, with the other set at the high limit. Relays $K1$ and $K2$ are operated by closure and opening of the meter contacts.

Relays $K1$ and $K2$ are connected so that, with no current flowing through their coils, contacts 1 and 3 are closed and contacts 2 and 3 are open. When, as illustrated, the liquid temperature drops to the point where the input signal from the thermocouple reaches the lower limit, the meter pointer touches the lower contact, energizing relay $K1$. As a result, the $K1$ contact 3 is separated from contact 1. Relay $K1$ contacts 2 and 3 then close,

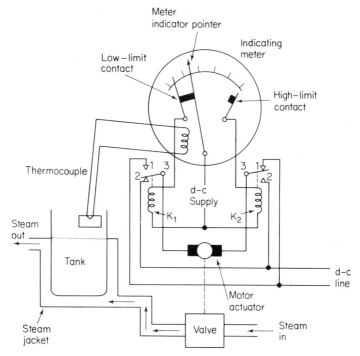

Figure 10-13 Elements of a basic floating control system used to control temperature of liquid in a tank.

and current flows through the d-c motor actuator. The actuator rotates in a direction to open the valve, permitting steam to enter the jacket. As a result, the liquid temperature rises.

When the liquid temperature starts to rise, the thermocouple output is increased, the meter contacts are separated, and relay *K*1 is deenergized. The motor-actuator stops rotating, but the valve remains in the open position. When the liquid temperature (and the input signal from the thermocouple) reaches the upper limit, the meter pointer touches the upper-limit contact, energizing relay *K*2. Now the motor-actuator rotates in the opposite direction, closing the steam valve. As the temperature starts to drop, the meter contacts separate, stopping rotation of the motor. The valve remains shut until the temperature drops to the point where *K*1 is energized once again.

With this type of floating control, the valve is actuated only when the movable contact of the meter touches either the lower or upper fixed contact. During the period when the movable contact is traveling between the two fixed contacts, the valve remains in the condition as last actuated.

10-4.2 *Pneumatic Controllers*

Pneumatic controllers generally operate by either of two basic principles. One system uses *force balance* (similar to null balance); the other system uses a form of *flapper nozzle* (similar to the pneumatic transmitter described in Sec. 10-1.2). The force-balance system is described in the following paragraphs.

ON-OFF force-balance system.   Figure 10-14 shows the basic principles of an ON-OFF force-balance pneumatic controller. For such a device, two pressure signals are balanced against each other. One signal is produced by the process variable acting through a primary sensing element and a transmitter. The other signal is a set-point pressure signal. Any deviation between the two pressures produces an error signal which, when amplified, produces an output pressure signal that is applied to the final control element. The control element then acts upon the the process variable to return the value to that determined by the set point, thus eliminating the deviation. The input signal, the set-point signal, and the output signal are all within a 3- to 15-psi range.

The controller shown in Fig. 10-14 consists of two sections. One section is an *unbalance detector,* which balances the input and set-point pressures against each other and detects the amount of deviation or error. The other section is an *amplifier,* which amplifies the error signal and produces the output pressure signal that is applied to the final control element.

The unbalance detector section is divided into four chambers by three flexible diaphragms. Diaphragms 1 and 3 have the same area. The area of

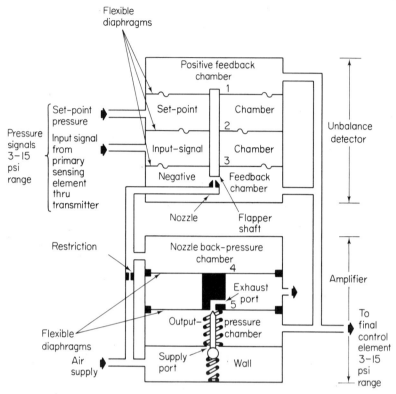

Figure 10-14 Pneumatic controller using ON-OFF force balance system.

diaphragm 2 is about half that of either diaphragms 1 or 3. The three diaphragms are connected by a shaft, the end of which acts as a flapper for a nozzle located in the negative feedback chamber.

The amplifier section is divided into four chambers by two flexible diaphrams and a wall. The two diaphragms (4 and 5) are coupled at their centers by the exhaust port. The pointed end of the step of the spring-loaded seat valve seats into the exhaust port. The ball end of the stem seats into the supply port in the wall separating the two lower chambers.

The input pressure signal from the transmitter is applied to the input-signal chamber. The set-point pressure signal, which may be adjusted manually, is applied to the set-point chamber. Note that the pressure in the positive-feedback chamber, exerted downward on diaphragm 1, is balanced by the pressure in the negative-feedback chamber, exerted upward on diaphragm 3, since both chambers obtain their pressures from the same source (the output pressure signal). Thus, if the input pressure and the set-point pressure are equal, the entire detector section is in balance and there is no resultant motion of the flapper shaft.

Should the input pressure rise as a result of an increase in the value of the process variable, the increased pressure pushes upward on diaphragm 2 and downward on diaphragm 3. However, since diaphragm 3 has a greater area then diaphragm 2, the net result is a downward movement of the flapper shaft, which brings the end of the shaft nearer the nozzle. As a result, the pressure of the air between the restriction and the nozzle is increased, causing an increased pressure in the nozzle back-pressure chamber of the amplifier section. The increased pressure forces diaphragms 4 and 5 downward against action of the spring, closing the exhaust port and forcing the valve stem downward, thus opening the supply port and permitting the supply air to enter the output-pressure chamber.

The output pressure continues to build up until the upward pressure on the bottom of diaphragm 5 becomes equal to the downward pressure on the top of diaphragm 4. At this point, the valve stem returns to its original position, with the supply of air to the output-pressure chamber shut off. Thus, the entire system becomes stabilized with a 15 psi output signal being applied to the final control element. Should the input pressure fall, the flapper shaft is moved upward, increasing the distance between the flapper and nozzle. The pressure in the nozzle back-pressure chamber is reduced, and diaphragms 4 and 5 are pushed up by the spring. The supply port is shut, and the exhaust port is opened. As a result, some of the air in the line to the final control element exhausts to the outside.

When enough air is exhausted so that pressure in the output pressure chamber is equal to the pressure in the nozzle back-pressure chamber, the exhaust port is closed. Now the entire system becomes stabilized, with a 3 psi output signal being applied to the final control element. If the pressure of the input signal increases, the flapper is moved closer to the nozzle and the pressure of the output signal is increased. This increases pressure in the positive-feedback chamber, a pressure which tends to move the flapper closer to the nozzle. Thus, the positive feedback pressure tends to aid the input signal.

On the other hand, the increases signal-output pressure also increases the negative-feedback pressure, which tends to move the flapper away from the nozzle. Thus, the negative-feedback pressure tends to oppose the input signal. However, since the positive- and negative-feedback pressures are equal, they cancel each other. A change in the pressure of the input signal of about 1% above the set-point pressure to 1% below the set-point pressure is sufficient to produce a full-range change in the output pressure to the final control element. That is, the controller changes from ON to OFF, or vice versa, very quickly.

The ON-OFF controller can be modified slightly to provide *proportional control,* where the final control element may assume a number of positions between the extremes, depending on the pressure of the input signal. This means that the output signal from the controller does not change quickly from 3 to 15 psi. Instead, the output pressure is able to assume intermediate values in response to the input signal.

10-5   READOUTS AND INDICATORS

In addition to meters and gages, the 7-segment digital readouts described in Sec. 8-6.8 have become quite popular in modern control and instrumentation systems. As mentioned such 7-segment readouts are generally part of a *display system.* The display system includes a *decoder,* which converts electrical data (representing the value to be displayed) into a turn-on of corresponding segments. In addition to decoders, the display system often includes drivers, counters, latches, and multiplexers (usually in IC form). Thus, before going into descriptions of the five most common types of 7-segment readouts, let us consider the basic elements of a display system.

10-5.1   7-Segment Display Systems

There are two basic types of display systems: *direct drive* and *multiplex.* Generally, it is more economical to multiplex displays of greater than four digits.

Direct drive displays.   The simplest type of display system, a clock counter shown in Fig. 10-15(a), consists of the four lines of BCD information feeding a decoder/driver. In turn, the decoder/driver turns on the corresponding segments of the 7-segment display. This direct drive system does not have information storage capability and thus reads out in real time.

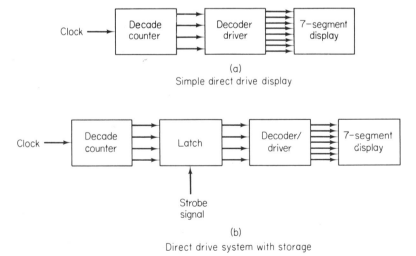

Figure 10-15 Basic direct-drive display and direct-drive display system with storage.

Another display system, one that is particularly used for less than four digits, is shown in Fig. 10-15(b). This system contains a decade counter, latch, decoder/driver, and display (one channel for every digit). The system has storage capability (the latches) which allows the counter to recount during the storage time. The latches hold the count until a "strobe" signal is applied.

Multiplex displays. The most commonly used system for multidigit displays is the multiplexed (or *time-shared* or *strobed*) system as shown in Fig. 10-16. By time-sharing one decoder/driver, the parts count, interconnections, and power can be saved. A multiplexer is a form of data selector which selects data on one or more input lines and applies the data to a single output channel, in accordance with a binary code applied to the control lines. The decade counters and latches (one for each digit) feed a scanned multiplexer whose sequenced BCD output drives *like segments* of the

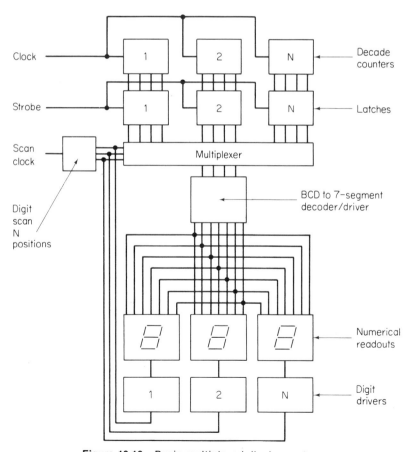

Figure 10-16 Basic multiplexed display system.

display. The digit select elements (digit drivers) are sequentially driven by the scan circuit which also synchronously drives the multiplexer. Thus, each display is scanned or strobed (turned on) in synchronism with the BCD data presented to the decoder. The display is scanned at a rate of 50 scans per second or faster, so that the display appears as continuously energized.

10-5.2 Liquid Crystal Displays (LCD)

Figure 10-17(a) shows the symbol for an LCD and the segment arrangement. Liquid crystals are fluids that flow like a liquid but which have some of the optical characteristics of solid crystals. LCD's consist of certain organic compounds whose characteristics change state when placed in an electric field. Thus, images can be created according to predetermined patterns (segments in this instance). Since no light is emitted or generated, very little power is required to operate LCD's and they are well suited for battery operation. LCD's have good readability in sunlight or bright light. However, for low ambient light conditions some form of light source (either within or external to the display) is generally used.

10-5.3 Fluorescent Displays

Figure 10-17(b) shows the symbol for fluorescent displays and the segment arrangement. Fluorescent displays, both diode and triode types, are similar electrically to vacuum-tube diodes and triodes. A difference is that the anodes of fluorescent display tubes are coated with a phosphor. When a positive voltage is placed across the anode and the heated cathode (or filament), electrons striking the anode cause the phosphor to fluoresce and emit light.

Triode fluorescent displays (with control grids) are somewhat easier to multiplex since the strobing is performed at a lower power level. Diode fluorescent displays are multiplexed in a relatively high power filament circuit. The display digits (segments) can be packaged in individual tubes or in a *planar multidigit* display contained in a single envelope with all like segments interconnected.

10-5.4 Light-Emitting Diodes (LED's)

Figure 10-17(c) shows the symbol for LED displays and the segment arrangement. LED's are semiconductor PN junction diodes (Chapter 7) which produce light when forward biased (anode more positive than cathode). The semiconductor material is either gallium arsenide phosphide (GaAsP) or gallimum phosphide (GaP), the former being more prevalent in red display applications. Similar to any junction diode, the voltage drop across the forward-biased junction is relatively constant. As shown in Fig. 10-17(c), the LED's (one for each segment) can be connected in a common-cathode or common-anode arrangement.

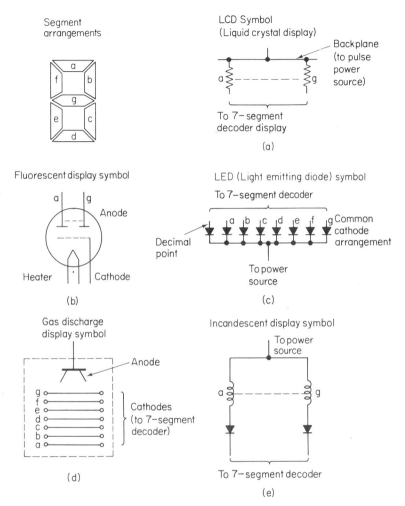

Figure 10-17 Segment arrangements and symbols for typical 7-segment readouts and indicators.

10-5.5 Gas Discharge Display

Figure 10-17(d) shows the symbol for gas discharge displays and the segment arrangement. First of the gas discharge tubes to reach the market and still in use with little noticeable change since their introduction in the mid 1950's is the *Nixie*® *tube*. They are a highly readable device, with a drawback of not having the numerals in one plane. The preshaped cathodes (0 to 9) are physically stacked within the envelope, which causes the display to jump in and out (or jitter) as the digits are changed, limiting the useful angle of view.

The more recently introduced *planar gas discharge* displays have their numerals in one plane, facilitating wide viewing angles, from 1½ to 16 digits being contained in one neon-filled envelope. Each digit has one anode and seven (or more) cathode segments. The like segments can be tied together as in most multidigit displays or can be brought out individually (if the number of digits is small).

When a voltage greater than the ionization potential (typically 170 V) is applied between the selected anode and cathode, the neon gas ionizes, and an orange glow appears around the cathode. For multiplexed displays, a blanking period is required between the cathode select and the anode scan pulses, ensuring that the previous digit is completely deionized before the following digit is strobed, thus preventing erroneous readouts.

10-5.6   Incandescent Display

Figure 10-17(c) shows the symbol for incandescent displays and the segment arrangement. Such displays have seven helical coil segments fashioned from tungsten alloy. When power is applied to each coil (or filament), the segment lights up, as is the case for any incandescent light. Power requirements are about the same as for LED's. The blocking diodes (one for each segment) prevent erroneous display indications through sneak electrical paths.

INDEX